# Pitman Research Notes in Mathematics Series

**Submission of proposals for consideration**
Suggestions for publication, in the form of outlines and representative samples, are invited by the Editorial Board for assessment. Intending authors should approach one of the main editors or another member of the Editorial Board, citing the relevant AMS subject classifications. Alternatively, outlines may be sent directly to the publisher's offices. Refereeing is by members of the board and other mathematical authorities in the topic concerned, throughout the world.

**Preparation of accepted manuscripts**
On acceptance of a proposal, the publisher will supply full instructions for the preparation of manuscripts in a form suitable for direct photo-lithographic reproduction. Specially printed grid sheets can be provided and a contribution is offered by the publisher towards the cost of typing. Word processor output, subject to the publisher's approval, is also acceptable.

Illustrations should be prepared by the authors, ready for direct reproduction without further improvement. The use of hand-drawn symbols should be avoided wherever possible, in order to maintain maximum clarity of the text.

The publisher will be pleased to give any guidance necessary during the preparation of a typescript, and will be happy to answer any queries.

**Important note**
In order to avoid later retyping, intending authors are strongly urged not to begin final preparation of a typescript before receiving the publisher's guidelines. In this way it is hoped to preserve the uniform appearance of the series.

**Longman Scientific & Technical**
**Longman House**
**Burnt Mill**
**Harlow, Essex, CM20 2JE**
**UK**
**(Telephone (0279) 426721)**

# John M Chadam

McMaster University, Canada

and

# Henning Rasmussen

University of Western Ontario, Canada

(Editors)

---

# Emerging applications in free boundary problems

## Proceedings of the International Colloquium 'Free Boundary Problems: Theory and Applications'

Longman Scientific & Technical

Copublished in the United States with
*John Wiley & Sons, Inc., New York*

CRC Press
Taylor & Francis Group
Boca Raton London New York

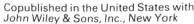

CRC Press is an imprint of the
Taylor & Francis Group, an **informa** business

A CHAPMAN & HALL BOOK

CRC Press
Taylor & Francis Group
6000 Broken Sound Parkway NW, Suite 300
Boca Raton, FL 33487-2742

First issued in hardback 2017

© 1993 by Taylor & Francis Group, LLC
CRC Press is an imprint of Taylor & Francis Group, an Informa business

No claim to original U.S. Government works

ISBN 13: 978-1-138-41755-7 (hbk)
ISBN 13: 978-0-582-08768-2 (pbk)

**Visit the Taylor & Francis Web site at**
**http://www.taylorandfrancis.com**

**and the CRC Press Web site at**
**http://www.crcpress.com**

**British Library Cataloguing in Publication Data**

A catalogue record for this book is
available from the British Library

**Library of Congress Cataloging-in-Publication Data**

A catalog record for this book is available

# Contents

## Chemical and biological reactions I

## Chemical and biological reactions II

## Control and identification

## Electromagnetism and electronics

# Preface

This is the first of three volumes containing the proceedings of the International Colloquium "Free Boundary Problems: Theory and Applications", held in Montreal (Canada) from June 13 to June 22, 1990.

The Scientific Committee was composed of J. Chadam, F. Clarke, A. Friedman, P. Fife, H. Glicksman, H. Rasmussen and I. Stakgold.

The Organizing Committee consisted of J. Chadam, F. Clarke, M. Delfour, M. Goldstein, B. Ladanyi and H. Rasmussen. Invaluable assistance and direction was received from the International Committee composed of A. Fasano, M. Frémond, K.-H. Hoffmann, M. Niezgódka, J. Ockendon, M. Primicerio and J. Sprekels.

The Montreal meeting was the fifth in a series of International Colloquia (Durham (UK) 1978, Montecatini (Italy) 1981, Maubuison (France) 1984, Irsee (Germany) 1987). The fruitful exchange which characterized these previous meetings continued at the Montreal meeting. Pure and applied mathematicians, applied scientists and engineers gathered to share their results on free boundary problems arising in their disparate disciplines. A wide spectrum of physical, numerical and mathematical methods were added to the common pool of knowledge in the subject. There were approximately 175 individuals from 25 countries whose energetic participation was the key to the success of the meeting. We are grateful to them all.

We are pleased to have been offered the privilege of holding the meeting for the first time in North America on the 100th anniversary of Stefan's work on the formation of ice in polar seas. In his paper, published a year later in 1891, he compared the solution (Lamé and Clapeyron (1831), Neumann (1860)) of the free boundary problem he posed with measurements obtained from polar explorations in the Canadian arctic. Presumably these data were obtained from the logs of explorers searching for a northwest passage during 1829-1853 (J.C. and H.R. thank C. Vuik for sharing this historical information from his thesis).

Financial support was received from École Polytechnique (Québec), FCAR (Québec), NASA (USA), NSERC (Canada) and Université de Montréal (Québec). We are grateful to M. Glicksman for organizing a wonderful session on microgravity sponsored entirely by NASA (USA). Finally we thank the director, F. Clarke, and the members and staff of the Centre de Recherches Mathématiques (Montréal) for providing the major part of the funding for the meeting and for their kind hospitality and assistance during our stay in Montreal. We shall be forever indebted to S. Chênevert, C. Doré and J. Roy for facilitating every aspect of the organization of the meeting with professional efficiency and charm.

In addition to the invited addresses of the main speakers, this volume contains the presentations in several non-traditional disciplines which are increasingly a source of interesting new free boundary problems. The papers included in the special session on Microgravity deal with flow and surface tension in low gravity. Mathematical modelling of Chemical and Biological Reactions is another rapidly advancing area giving rise to many new types of free boundary problems including motion by mean curvature. This volume contains a wide spectrum of papers in this area covering applications from combustion in porous media to controlled release of drugs to osmosis. By applying the techniques of optimal control to free boundary problems a new important class of problems is becoming tractable. Some of the theoretical results in this approach as well as some applications to jets and shape memory alloys appear in this volume. Finally, the last section deals with free boundary problems from the semiconductor industry.

Hamilton and London, Canada
November, 1992

J. Chadam
H. Rasmussen

# Main speakers

M CHIPOT

# New remarks on the dam problem

## 0. Introduction

Let $\Omega$ be a bounded Lipschitz domain in $\mathbf{R}^2$. $\Omega$ represents the section of a porous medium. The points in $\mathbf{R}^2$ will be denoted by $(x, y)$.

The boundary $\Gamma$ of $\Omega$ is divided into three parts: an impervious part $S_1$, a part in contact with the air $S_2$ and finally a part covered by fluid $S_3$ (see figure 1).

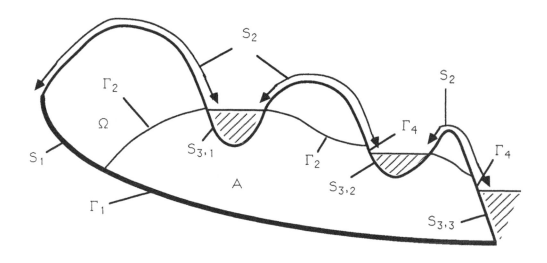

Figure 1

For convenience, we will assume that $S_1$, $S_3$ are relatively open in $\Gamma$ and we will denote by $S_{3,i}$, $i = 1, \ldots, N$ the different connected components of $S_3$.

Assuming that the flow in $\Omega$ has reached a steady state we are concerned with finding the pressure $p$ of the fluid and the part of the porous medium where some flow occurs -i.e. the wet set $A$.

Note that this two dimensional model could describe for instance the steady flow in the cross section $\Omega$ of a longitudinal porous medium.

This problem has been studied intensively in the last two decades. The initial contributions are due in particular to the Pavia school (see [Ba.$_1$]-[Ba.$_3$], [B.C.], [V.]).

## 1. Mathematical setting

The boundary of $A$, that we will denote by $\partial A$, is divided into four parts: an impervious part $\Gamma_1$, a free boundary $\Gamma_2$, a part covered by the fluid $\Gamma_3 = S_3$ and finally a seepage front $\Gamma_4$ where the fluid is flowing out $\Omega$ but does not remain in a significant amount to cover the porous medium.

The velocity $v$ of the fluid in $A$ is given, according to Darcy's law, by

$$v = -k\nabla(p+y) \tag{1.0}$$

where $p$ is the pressure, $k$ is the coefficient of permeability of the medium. From now on, we will assume that everything has been scaled in such a way that $k = 1$.

Then, assuming the fluid incompressible we have

$$\text{div } v = 0 \quad \text{in} \quad A$$

or

$$\Delta p = 0 \quad \text{in} \quad A. \tag{1.1}$$

Next, on $\Gamma_1 \cup \Gamma_2$ there is no flux of fluid through this part of boundary. So, if $\nu$ denotes the outward unit normal to $\partial A$ one has

$$v.\nu = 0 \quad \text{on} \quad \Gamma_1 \cup \Gamma_2$$

or by (1.0)

$$\frac{\partial}{\partial \nu}(p+y) = 0 \quad \text{on} \quad \Gamma_1 \cup \Gamma_2. \tag{1.2}$$

On $\Gamma_4$ the fluid is free to exit from the porous medium and thus one has

$$v \cdot \nu \geq 0 \quad \text{on} \quad \Gamma_4$$

which reads again by (1.0)

$$\frac{\partial}{\partial \nu}(p + y) \leq 0 \quad \text{on} \quad \Gamma_4. \tag{1.3}$$

We will denote by $\varphi$ the outside pressure on $S_2 \cup S_3$. For instance, if we assume that we are in the case of the figure 1, and if the atmospheric pressure has been scaled to 0, then $\varphi$ is given by

$$\varphi(x, y) = 0 \quad \text{if} \quad (x, y) \in S_2$$

$$= h_i - y \quad \text{if} \quad (x, y) \in S_{3,i}, \quad i = 1, \ldots, N$$

where $h_i$ denotes the level of the reservoir covering $S_{3,i}$ -we assume that the fluid is water and the units chosen in such a way that the hydrostatic pressure of a reservoir of level $h$ is given by $h - y$. Clearly $\varphi$ is a Lipschitz continuous function.

Besides (1.2) there is a second condition on $\Gamma_2$ –i.e. the pressure $p$ coincides with the atmospheric pressure in such a way that

$$p = 0 \quad \text{on} \quad \Gamma_2. \tag{1.4}$$

In [C.C.$_1$] or [C.] we studied the case of Dirichlet boundary conditions -i.e. the case where $p$ is prescribed equal to $\varphi$ on $S_3$. We would like to consider here a model where the flux is prescribed on this part of the boundary. More precisely we will assume that the flux is governed by a function of the jump of pressure across $S_3$. This kind of conditions is called a "leaky boundary condition" and we refer the reader to [Be.] for physical justifications of such a model (see also [R] for the relationship between the two models).

So, we will impose

$$\frac{\partial}{\partial \nu}(p + y) = \beta(x, \varphi - p) \quad \text{on} \quad S_3. \tag{1.5}$$

On $\beta$ we will assume for instance

$$\beta(x, 0) \in L^2(S_3) \tag{1.6}$$

(see for instance [N.] for the meaning of this space, note that $x$ denotes here and in the following a point in $\mathbf{R}^2$ and not its first entry) and also

$$x \to \beta(x, u) \quad \text{is measurable for every} \quad u \in \mathbf{R}, \tag{1.7}$$

$\exists C > 0$ such that

$$|\beta(x, u_1) - \beta(x, u_2)| \leq C|u_1 - u_2| \quad \text{a.e.} \quad x \in S_3, \quad \forall u_1, u_2 \in \mathbf{R}, \tag{1.8}$$

$$u \to \beta(x, u) \quad \text{is nondecreasing for a.e.} \quad x \in S_3, \tag{1.9}$$

$$\beta(x, u) \geq 0 \quad \forall u \geq 0 \quad \text{a.e.} \quad x \in S_3. \tag{1.10}$$

So, assuming that everything is smooth the problem is to find a pair $(p, A)$ such that (1.1)-(1.5) holds. This is the strong formulation. First remark that to find the pair $(p, A)$ is equivalent to find the pair $(p, \chi_A)$ where $\chi_A$ denotes the characteristic function of the set $A$. Then, following [B.K.S.], (see also [A.]), it is easy to see that the problem (1.1)-(1.5) can be recasted into a weak form which is:

Find $(p, \chi) \in H^1(\Omega) \times L^\infty(\Omega)$ such that

(i)  $p \geq 0$, $0 \leq \chi \leq 1$  a.e. in $\Omega$, $\chi = 1$ on $[p > 0]$,

(ii)  $p = 0$  on  $S_2$, $\tag{P}$

(iii)  $\displaystyle\int_\Omega \nabla p \cdot \nabla \xi + \chi \xi_y \, dx - \int_{S_3} \beta(x, \varphi - p) \cdot \xi \, d\sigma(x) \leq 0 \quad \forall \xi \in H^1(\Omega), \ \xi \geq 0 \text{ on } S_2.$

($[p > 0] = \{(x, y) \in \Omega \mid p(x, y) > 0\}$, $d\sigma(x)$ denotes the superficial measure on $S_3$, see [K.S.], [G.T.] for the definition of $H^1(\Omega)$). We will refer to $(P)$ as the weak formulation of our initial problem. Clearly if (1.1)-(1.5) has a solution $(p, A)$ and if $p$ denotes also the extension of $p$ by 0 outside $A$ then it is easy to show that $(p, \chi_A)$ is a solution to $(P)$ and thus any strong solution to (1.1)-(1.5) will be found among those of $(P)$.

## 2. Existence and first properties of the solutions

First we can show that the problem (P) admits a solution. Indeed we have

**THEOREM 1:** Assume that $\beta$ is a function satisfying (1.6)-(1.10) , then there exists a solution $(p, \chi)$ to the problem $(P)$.

**Proof** : The idea is the one introduced in [B.K.S.] (see also [A.]). First one considers the approximated problem: find $p_\varepsilon \in H^1(\Omega)$ such that $p_\varepsilon = 0$ on $S_2$ and

$$\int_\Omega \nabla p_\varepsilon \nabla \xi + H_\varepsilon(p_\varepsilon) \cdot \xi_y \, dx - \int_{S_3} \beta(x, \varphi - p_\varepsilon).\xi \, d\sigma(x) = 0 \ \forall \, \xi \in H^1(\Omega), \ \xi = 0 \text{ on } S_2 \ (P_\varepsilon)$$

$H_\varepsilon$ is the approximation of the Heaviside graph defined by

$$H_\varepsilon(p) = 0 \vee \frac{1}{\varepsilon} p \wedge 1 \tag{2.1}$$

where $\vee$ denotes the maximum of two functions and $\wedge$ the minimum, $\varepsilon$ is positive. Using the Schauder fixed point Theorem it is easy to show that the problem $(P_\varepsilon)$ admits a unique solution. Moreover one has

$$0 \leq p_\varepsilon \quad \text{in } \Omega.$$

Then, one proves that $p_\varepsilon$ is bounded in $H^1(\Omega)$ and one extracts from $(p_\varepsilon, H_\varepsilon(p_\varepsilon))$ a subsequence that converges toward $(p, \chi)$ that satisfies (P). We refer the interested reader to [B.K.S.], [C.C.$_1$], [C.C.$_2$], [C.] for details.

**Remark 1** : It is possible to prove existence of a solution to (P) under some other assumptions on $\beta$. (see [C.C.$_2$]).

The next interesting question is to decide if the solution to (P) is unique. This problem remains widely open in the case of a general dam. We would like in the following of this note stress out the differences with our model and the classical one where Dirichlet boundary conditions are imposed. First we have:

**PROPOSITION 1:** Let $(p, \chi)$ be a pair solution to $(P)$. Then one has in the distributional sense.

$$\Delta p + \chi_y = 0 \quad \text{in} \quad \Omega \tag{2.2}$$
$$\Delta p \geq 0 \quad , \quad \chi_y \leq 0 \quad \text{in} \quad \Omega. \tag{2.3}$$

**Proof:**  This follows [C.C.$_1$] or [C.].

As an obvious consequence since $\chi_y \in W^{-1,s}(\Omega)$ for any $s$ (see [N.] for a definition of this space) we have:

6

**COROLLARY 1:** Let $(p, \chi)$ a pair of solution to $(P)$ then for any $s > 1$

$$p \in W_{loc}^{1,s}(\Omega).$$

Then we can show:

**PROPOSITION 2:** Let $(p, \chi)$ be a solution to $(P)$. Let $(x_0, y_0) \in \Omega$. If $p(x_0, y_0) > 0$ then there exists $\varepsilon > 0$ such that the cylinder

$$C_\varepsilon = \{(x, y) \in \Omega \mid |x - x_0| < \varepsilon , \ y < y_0 + \varepsilon\}$$

lies in the set $[p > 0]$.

If $p(x_0, y_0) = 0$ then $p(x_0, y) = 0 \ \ \forall \ (x_0, y) \in \Omega, \ y > y_0$.

**Proof:** Note that due to corollary 1 $[p > 0]$ is an open subset of $\Omega$. Then it is enough to combine Proposition 1 and the maximum principle to conclude.

The physical meaning of Proposition 2 is that as soon as a point $(x_0, y_0)$ is saturated -i.e. $p((x_0, y_0)) > 0$- then all the dam below it is saturated. In the case of Dirichlet boundary conditions it can be shown that the porous medium is saturated below $S_3$ due to the fact that $p = \varphi > 0$ on $S_3$. This is no more the case here and one can have $p = 0$ below $S_3$. This is the case for instance when $\beta$ is small. Indeed in this case the flux through $S_3$ is too weak to saturate the dam (see [C.C.$_2$]). We will see in the next section an example of this phenomenon (see Proposition 4). Before we would like to check that our model leads to a free boundary problem -i.e. that the set $[p = 0]$ has a positive measure. This is done through the following proposition:

**PROPOSITION 3:** Let us assume that $\beta$ satisfies (1.7)-(1.10) and

$$\beta(x, u) \cdot u \geq 0 \ \ \forall \, u \in \mathbf{R}, \ \text{a.e.} \ x \in S_3. \tag{2.4}$$

Let $(p, \chi)$ be a solution to $(P)$ and let us denote by $h_1$ the level of the highest reservoir. Then one has

$$(p, \chi) = (k - y, 1) \tag{2.5}$$

on any connected component of $[p > 0]$ that intersects $[y > h_1]$. Moreover, outside these connected components one has

$$p \leq (h_1 - y)^+. \tag{2.6}$$

**Proof:** One inserts

$$\xi = (p - (h_1 - y)^+)^+$$

in $(P)$ (iii) and the result follows after some arguments similar to those of $[C.C._1]$.

**Remark 2:** As in $[C.C._1]$ let us call a "pool" a function like in (2.5). Clearly when one cannot have a pool above the highest level, then $(p, \chi) = (0, 0)$ there and the problem has a free boundary.

On the other hand, when the geometry of the dam allows the occurence of a "pool" then nonuniqueness arises as in the case of Dirichlet boundary conditions (see $[C.C._1]$ or $[C.]$). We do not know at this time if uniqueness of a solution to $(P)$ holds modulo these "pool" functions.

### 3. A more striking behavior

An important issue, which plays an essential role in proving uniqueness (modulo "pools") in the case of Dirichlet boundary conditions is to show that $\chi$ is the characteristic function of the set $[p > 0]$. As we will see this is no more true for our model. Let us consider the case of the following figure where $S_1 = \emptyset$.

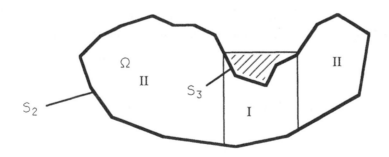

figure 2

First we would like to show that in this case when

$$0 \le \beta(\varphi)/\nu_y \le 1 \tag{3.1}$$

then the only solution to $(P)$ is given by

$$(p, \chi) = (0, (\beta(\varphi)/\nu_y)\chi_I)$$

where $\chi_I$ denotes the characteristic function of the region below $S_3$ and denoted by $I$ on the figure 2, $\nu_y$ is the $y$ entry of the unit outward normal to $\Gamma$ on $S_3$ (we have denoted by $\beta(\varphi)/\nu_y$ the function independent of $y$ equal to $\beta(\varphi)/\nu_y$ on $S_3$). Note that this is a particular case where uniqueness holds. Moreover, this shows that we cannot expect in general $\chi$ to be a characteristic function of a set ($\chi$ is not if $\beta(\varphi)/\nu_y < 1$). So, the situation is quite different of the one in the classical dam problem.

So, let us prove:

**PROPOSITION 4:** Assume that $\beta$ is independent of $x$, $\beta(0) = 0$, and satisfies (1.8), (1.9). Then, if (3.1) holds the problem $(P)$ corresponding to the figure 2 has a unique solution given by

$$(p, \chi) = (0, (\beta(\varphi)/\nu_y)\chi_I). \tag{3.2}$$

**Proof:** First let us check that $(p, \chi)$ given by (3.2) satisfies $(P)$. We only have to check $(P)$ (iii). For that note that since $\beta(\varphi)/\nu_y$ is a function of $x$ only

$$\int_\Omega \nabla p \cdot \nabla \xi + \chi \xi_y \, dx - \int_{S_3} \beta(\varphi - p) \cdot \xi \, d\sigma(x) = \int_I \frac{\beta(\varphi)}{\nu_y} \xi_y \, dx - \int_{S_3} \beta(\varphi) \cdot \xi \, d\sigma(x)$$

$$= \int_I \left( \frac{\beta(\varphi)}{\nu_y} \cdot \xi \right)_y dx - \int_{S_3} \beta(\varphi) \cdot \xi \, d\sigma(x) \tag{3.3}$$

$$= \int_{\partial I \setminus S_3} \frac{\beta(\varphi)}{\nu_y} \cdot n_y \cdot \xi \, d\sigma(x) \leq 0$$

for any $\xi \geq 0$ on $S_2$, $n_y$ denotes the $y$ entry of the outward unit normal to $\partial I \setminus S_3$ and thus $n_y \leq 0$ (see figure 2). $\partial I$ denotes the boundary of $I$. So, $(p, \chi)$ given by (3.2) is a solution to $(P)$. Note that from (3.3) one deduces easily that

$$\int_\Omega \chi \cdot \xi_y \, dx - \int_{S_3} \beta(\varphi)\xi \, d\sigma(x) = 0 \quad \forall\, \xi \in H^1(\Omega),\ \xi = 0 \ \text{on}\ \partial I \cap S_2. \tag{3.4}$$

Let us now denote by $(p', \chi')$ an other solution to (P). Thus, one has

$$\int_\Omega \nabla p' \cdot \nabla \xi + \chi' \xi_y \, dx - \int_{S_3} \beta(\varphi - p') \cdot \xi \, d\sigma(x) \leq 0 \quad \forall\, \xi \in H^1(\Omega),\ \xi \geq 0 \ \text{on}\ S_2. \tag{3.5}$$

Taking $\xi = -p'$ in (3.4) and $\xi = p'$ in (3.5) and adding one gets

$$\int_\Omega |\nabla p'|^2 + (\chi' - \chi) \cdot p'_y \, dx - \int_{S_3} \beta(\varphi - p') - \beta(\varphi) \cdot p' \, d\sigma(x) \leq 0. \tag{3.6}$$

9

But

$$\int_\Omega (\chi' - \chi) p'_y \, dx = \int_\Omega (1 - \chi) p'_y \, dx$$

$$= \int_I \left(1 - \frac{\beta(\varphi)}{\nu_y}\right) p'_y \, dx + \int_{II} p'_y \, dx$$

$$= \int_I \left(\left(1 - \frac{\beta(\varphi)}{\nu_y}\right) p'\right)_y \, dx$$

$$= \int_{S_3} \left(1 - \frac{\beta(\varphi)}{\nu_y}\right) \cdot \nu_y \cdot p' \, d\sigma(x) \geq 0.$$

(3.7)

Moreover, from the fact that $\beta$ is nondecreasing one has

$$- \int_{S_3} \beta(\varphi - p') - \beta(\varphi) \cdot p' \, d\sigma(x) \geq 0.$$

(3.8)

Combining (3.6), (3.7), (3.8) one deduces

$$\int_\Omega |\nabla p'|^2 \, dx \leq 0$$

and thus $p' = p = 0$. But now (3.5) reads

$$\int_\Omega \chi' \xi_y \, dx - \int_{S_3} \beta(\varphi) \cdot \xi \, d\sigma(x) \leq 0 \quad \forall \, \xi \in H^1(\Omega), \, \xi \geq 0 \quad \text{on} \ S_2.$$

(3.9)

Using (2.2) one deduces

$$\chi'_y = 0 \quad \text{or} \quad \chi' = \chi'(x).$$

Then if we denote by $T$ and $B$ respectively the upper and the bottom part of $\Gamma$ one deduces from (3.9) for $\xi \in H^1(\Omega), \, \xi \geq 0$ on $S_2$

$$0 \geq \int_\Omega (\chi'\xi)_y \, dx - \int_{S_3} \beta(\varphi) \cdot \xi \, d\sigma(x)$$

$$= \int_T \chi' \cdot \xi \cdot \nu_y \, d\sigma(x) + \int_B \chi' \cdot \xi \cdot \nu_y \, d\sigma(x) - \int_{S_3} \beta(\varphi) \cdot \xi \, d\sigma(x).$$

(3.10)

Taking in (3.10) any $\xi$ that vanishes on $S_2$ one gets

$$\int_{S_3} (\chi' \nu_y - \beta(\varphi)) \cdot \xi \, d\sigma(x) = 0$$

for such a $\xi$. Hence $\chi' = \beta(\varphi)/\check{\nu}_y$ on $S_3$. Then (3.10) becomes

$$0 \geq \int_{\Gamma \setminus S_3} \chi' \cdot \xi \cdot \nu_y \, d\sigma(x) \quad \forall \xi \in H^1(\Omega), \ \xi \geq 0 \quad \text{on} \ \ S_2.$$

Taking $\xi \geq 0$ on $T$, $\xi = 0$ on $B$ one deduces

$$0 \geq \int_{T \setminus S_3} \chi' \cdot \xi \cdot \nu_y \, d\sigma(x)$$

hence $\chi' = 0$ on $T \setminus S_3$ and the result follows. (We have assumed $\nu_y > 0$ on $T$). This completes the proof of the theorem.

**Remark 3:** The above example - which shows that $\chi$ is not necessarily a characteristic function - indicates that the techniques of [C.C.$_1$] to prove uniqueness do not apply here directly.

## REFERENCES

[A.] H.W. Alt: Strömungen durch inhomogene poröse Medien mit freiem Rand.
Journal für die reine und angewandte Mathematik 305, (1979), pp. 89–115.

[Ba.$_1$] C. Baiocchi: Su un problema di frontiera libera connesso a questioni di idraulica.
Ann. Mat. Pura Appl. 92, (1972), pp. 107–127.

[Ba.$_2$] C. Baiocchi: Free Boundary Problems in the Theory of Fluid Flow Through Porous Media.
Proceedings of the International Congress of Mathematicians, Vancouver, (1974), pp. 237–243.

[Ba.$_3$] C. Baiocchi: Free Boundary Problems in Fluid Flow Through Porous Media and Variational Inequalities.
in Free Boundary Problems, Proceedings of a seminar held in Pavia Sept. - Oct. 1979, Vol. 1, Rome, (1980), pp. 175–191.

[Be.] J. Bear: **Hydraulics of Groundwater**.
McGraw-Hill, (1979), New York.

[B.C.] C. Baiocchi - A. Cappello: **Disequazioni variazionali et quasivariazionali. Applicazioni a problemi di frontiera libera.**
Vol. 1 et 2, Pitagora Editrice, (1978), Bologna.

[B.K.S.] H. Brezis, D. Kinderlehrer and G. Stampacchia: Sur une nouvelle formulation du problème de l'écoulement à travers une digue.
C.R. Acad. Sc. Paris Série A 287, (1978), pp. 711–714.

[C.C.$_1$] J. Carrillo and M. Chipot: On the Dam Problem.
Journ. of Diff. Equ. 45, (1982), pp. 234–271.

[C.C.$_2$] J. Carrillo and M. Chipot: The Dam Problem with Leaky Boundary Condi tions.
(To appear).

[C.] M. Chipot: **Variational Inequalities and Flow in Porous Media.**
Applied Math. Sciences Series # 52, Springer Verlag, (1984), New York.

[G.T.] D. Gilbarg and N.S. Trudinger: **Elliptic Partial Differential Equations of Second Order.**
Springer, (1977), New York.

[K.S.] D. Kinderlehrer and G. Stampacchia: **An Introduction to Variational In equalities.**
Academic Press, (1980), New York.

[N.] J. Nečas. **Introduction to the Theory of Nonlinear Elliptic Equations.**
John Wiley Interscience, (1986), New York.

[R.] J.F. Rodrigues: On the dam problem with boundary leaky condition.
Portugaliae Mathematica, 39, (1980), pp. 399–411.

[V.] A. Visintin. Study of a Free Boundary Filtration Problem by a nonlinear Varia tional Equation.
Bollettino U.M.I. 5, (1979), pp. 212–237.

Université de Metz
Département de Mathématiques
Ile de Saulcy, 57045 Metz-Cedex 01
(FRANCE)

S HOWISON
# Codimension-two free boundary problems

### Abstract

Codimension-two free boundary problems are obtained by a linearisation process from normal free boundary problems whose free boundary lies close to and nearly parallel to a fixed boundary. We describe two examples, and we make some general conjectures on the relationship between a codimension-two problem and its prototype.

## 1    Introduction

The purpose of this article is to describe several free boundary problems which all have the property that the dimensionality of their free boundary is two fewer (not, as usual, one fewer) than the dimension of the space they occupy: hence the 'codimension-two' of the title. Such problems arise when the free boundary $\Gamma$ of a codimension-one free boundary problem lies near to and nearly parallel to a fixed or known surface; Figure 1 illustrates the geometry. An approximate solution in the 'thin' region II (for two-phase problems), combined with linearization of $\Gamma$ onto the segment $\Gamma'$ of the fixed surface, yields a new free boundary problem whose free boundary is just the intersection of the original free boundary with the fixed surface. In general, one hopes that the new problem will be easier to solve than the original, bearing in mind in particular the difficulties of resolving the free boundary, while still retaining its important qualitative and quantitative features.

Several problems of this kind have been studied, although not from the point of view of a general classification. We shall describe two, the 'gently-sloping dam' problem and the electropainting problem, and we shall briefly mention a number of others, concluding with some conjectures and open questions.

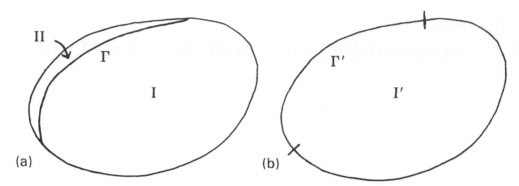

Figure 1: A free boundary problem and its codimension-2 version

This paper describes joint work with John Ockendon, and a parallel description of codimension-two classification, with different examples, is given by him in [Ock].

## 2  The gently-sloping dam problem

A review of the classical Dam Problem is given by Chipot in these Proceedings. Our interest is in situations where the top surface of the dam, $y = \varepsilon g(x)$ is nearly horizontal and gently-sloping, i.e. $\varepsilon \ll 1$, and the other boundary conditions are such that the free surface too lies near $y = 0$, as in Figure 2. The practical application of this configuration is described in [Aitch1], where the linearised model below is derived.

In two dimensions, the full problem for the dimensionless velocity potential $\Phi(x, y, t) = -(p + y)$, where $p$ is the pressure, is

$$\Delta \Phi = 0 \tag{1}$$

in the liquid region $\Omega$, with

$$p = 0, \quad \text{i.e.} \quad \Phi = -y \tag{2}$$

on free surfaces and seepage faces; also on the free surfaces,

$$\mathbf{n}.\nabla\Phi = V_n \tag{3}$$

where $V_n$ is the velocity in the direction of the normal $\mathbf{n}$, while on the seepage faces we have the outflow condition

$$\mathbf{n}.\nabla\Phi > 0. \tag{4}$$

We assume that $g(x)$ is bounded and sufficiently smooth, and we impose appropriate fixed boundary conditions, for example no flow across a lower border $y = -H$; for simplicity we only consider domains for which $-\infty < x < \infty$.

Exploiting the fact that $\varepsilon \ll 1$, we seek a solution in which the free surface is $y = \varepsilon h(x, t)$, and then writing $\Phi = \varepsilon\phi(x, y, t)$ and linearising the free surface and seepage face conditions onto $y = 0$, we obtain the problem

$$\Delta\phi = 0 \quad y < 0 \tag{5}$$

with

$$\left. \begin{aligned} \phi &= -g(x) \\ \partial\phi/\partial y &\geq 0 \end{aligned} \right\} \tag{6}$$

on projections of seepage faces onto $y = 0$, while

$$\left. \begin{aligned} \phi &= -h(x, t) \\ \frac{\partial\phi}{\partial y} &= \frac{\partial h}{\partial t} \end{aligned} \right\} \tag{7}$$

on projection of free surfaces onto $y = 0$ (ie $\Gamma'$). The other fixed boundary conditions remain as before.

The points where the boundary conditions switch from (6) to (7) are now the only 'free boundaries' of the problem; we call them the *codimension-two free boundary* of the linearized problem, and there we assume that

$$|\nabla\phi| < \infty \tag{8}$$

which is conjectured to determine the codimension-two free boundary uniquely for this problem.

For certain fixed boundary conditions, (5)–(8) can be reformulated as an evolutionary variational inequality, and existence, uniqueness and monotonicity of the codimension-two free boundary in this special case are proved in [Ell].

However, several questions remain open. For example, the codimension-two free boundary may advance towards the seepage face (as in [Ell]), or it may retreat as the water rises, driven by fixed boundary conditions below, on $y = -H$. Is there any substantive difference between those two cases? Considering the obvious 3-dimensional generalization, is the codimension-two free boundary stable or unstable to small two-dimensional perturbations in the plane $y = 0$, and again is there any difference between advancing and receding? We recall that in the full problem a free boundary whose normal points upwards is linearly stable unless its vertical velocity component

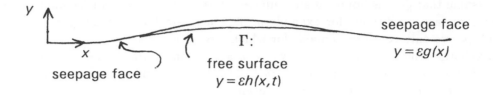

$$\Delta \Phi = 0$$

$$y = -H$$

Figure 2: The gently sloping dam problem

is less than $-1$ in dimensionless terms (ie the fluid is being sucked down faster than it would fall under gravity alone). It would be interesting to see whether an analogous result holds for the approximate version too.

We make two remarks, concerning the limits $H \to \infty$, $H \to 0$. In the former case, (5)-(8) can be transformed into an integro-differential equation as follows. Assume for simplicity that the codimension-two free boundary consists of the two points $x = \pm d(t)$, with symmetry about $x = 0$, and with seepage faces in $|x| < d(t)$ (Figure 2 with $H = \infty$). Let $\phi_g$ be the solution to $\Delta \phi_g = 0$ in $y < 0$, with $\phi_g(x, 0) = -g(x)$ and $|\nabla \phi_g| \to 0$ as $y \to -\infty$. Then $\tilde{\phi} = \phi - \phi_g$ is harmonic in $y < 0$ and vanishes on $y = 0$ for $|x| > d(t)$. Thus

$$\frac{\partial \tilde{\phi}}{\partial y}(x, 0, t) = -\frac{1}{\pi} \fint_{-\infty}^{\infty} \frac{\partial \tilde{\phi}}{\partial \xi}(\xi, 0, t) \frac{d\xi}{\xi - x}$$

$$= -\frac{1}{\pi} \fint_{-d(t)}^{d(t)} \left[ g'(\xi) - \frac{\partial h}{\partial \xi}(\xi, t) \right] \frac{d\xi}{\xi - x}.$$

But for $|x| < d(t)$ and $y = 0$, $\partial \tilde{\phi} / \partial y = \partial h / \partial t - \partial \phi_g / \partial y$. Therefore, for $|x| < d(t)$,

$$\frac{\partial h}{\partial t}(x, t) = \frac{1}{\pi} \fint_{-d(t)}^{d(t)} \frac{\partial h}{\partial \xi}(\xi, t) \frac{d\xi}{\xi - x} + \frac{\partial \phi_g}{\partial y}(x, 0) - \frac{1}{\pi} \fint_{-d(t)}^{d(t)} \frac{g'(\xi) d\xi}{\xi - x} \qquad (9)$$

with

16

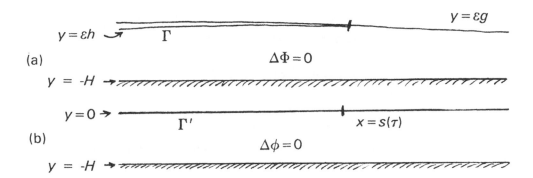

Figure 3: 'Long and thin' gently-sloping dam

$$h(x,0) = h_0(x) \leq g(x). \tag{10}$$

In the simple case $g(x) = 0$, an explicit solution [Ock] is available:

$$h(x,t) = \frac{t}{\pi} \int_{-d(0)}^{d(0)} \frac{h_0(\xi)d\xi}{t^2 + (x - \xi)^2}.$$

In this case the free boundary immediately disappears, with $h(x,t) < 0$ for all $t > 0$. A similar formula may be found from (9) when $g$ is not zero, but the inhomogeneous terms may mean that it violates the constraint $h(x,t) \leq g(x)$. In this case the free boundary must persist, and the singular integral equation is to be solved with a unilateral constraint.

Secondly, consider the limit $H \to 0$, but with $\varepsilon \ll H \ll 1$ so that our linearization is still valid. Assume for simplicity that $g'(x) < 0$ and that there is just one component $x = s(t)$ of the codimension-two free boundary as in Figure (3).

Write $y = HY$, and then since $\partial^2 \phi / \partial Y^2 + H^2 \partial^2 \phi / \partial x^2 = 0$,

$$\phi \sim \phi_0(x,t) - \frac{1}{2}H^2((Y + 1)^2 \partial^2 \phi_0(x,t)/\partial x^2 + \phi_2(x,t)) + o(H^2).$$

Now from (6,7),

$$\begin{aligned}
\phi_0 &= -h(x,t) & x < s(t) \\
&= -g(x) & x > s(t).
\end{aligned}$$

Equation (7) also shows that we must rescale $t$ by setting $t = H^{-1}\tau$, and then from (7)

$$\frac{\partial \phi_0}{\partial \tau} = \frac{\partial^2 \phi_0}{\partial x^2}, \quad x < s(\tau),$$

with $\phi(s(\tau), \tau) = -g(s(\tau))$. We also assume continuity of $\partial\phi/\partial x$ at $s(\tau)$, so that $\phi_x(s(\tau), \tau) = -g'(s(\tau))$. Putting $\psi = \phi_0 + g(x)$, we have

$$\psi_\tau = \psi_{xx} + g''(x), \quad x < s(\tau),$$

and

$$\psi = \psi_x = 0, \quad x \geq s(\tau).$$

Also $\psi \geq 0$ since to lowest order $h(x, t) = -\phi_0 \leq g(x)$, and so $\psi$ satisfies a version of the oxygen consumption problem [Crank] with "source" $g''(x)$. Thus if $h(x, 0) = g(x)$, the codimension-two free boundary immediately jumps to the inflexion point(s) of $g$. (It is an interesting conjecture that this also occurs when H is not small.) We also remark that we can make the codimension-two free boundary move in either direction by injecting or withdrawing small amounts of fluid at $y = -H$, and the indifference of the free boundary of the oxygen consumption problem to whether it advances or retreats is a small piece of evidence that the gently-sloping dam problem is well-posed both ways.

Finally, if $H \to 0$ with $\varepsilon \sim H \ll 1$, we retrieve the full Boussinesq/Dupuit model $(hh_x)_x = h_\tau$ with $h = g(x)$, $h_x = g'(x)$ on the free boundary $x = s(\tau)$; but this is no longer a codimension-two free boundary problem.

## 3  The electropainting problem

An effective way of painting a complex metal object is by electropainting. In this process, the workpiece is immersed in a tank of paint, and a potential difference is maintained between it and anodes around the sides of the tank. Positively charged paint particles migrate under the influence of this potential difference, and adhere to the metal, forming a thin layer of nearly solid paint. Two crucial features are (a) that paint will not adhere either to existing paint or to bare metal unless the current density exceeds a critical value, the *deposition current density*; and (b) that the resistivity of paint in solution is very much less than that of deposited paint. The former means that there may be bare patches, especially in shielded parts of the body, while the latter means that the paint layers are very thin.

A dimensionless model of the process is described by [Aitch2]. The electric potential $\Phi(\mathbf{x}, t)$ satisfies

$$\Delta\Phi = 0 \tag{11}$$

in the paint solution (region I) and deposited paint layers (region II). On the interface $\Gamma$ between I and II, where the paint layer has nonzero thickness,

$$[\Phi]_I^{II} = 0, \qquad \frac{\partial \Phi}{\partial n}\bigg|_I = \varepsilon \frac{\partial \Phi}{\partial n}\bigg|_{II} \tag{12}$$

where $\varepsilon \ll 1$ is the ratio of the resistivities in I and II; the second equation of (12) expresses continuity of current. In addition, the rate of growth of the paint is assumed to be proportional to the surface current density minus the deposition current density, so, after scaling,

$$V_n = \frac{\partial \Phi}{\partial n} + \delta \tag{13}$$

where $V_n$ is the normal velocity of $\Gamma$ and $\delta$ is the deposition current density (the signs in (13) take n as shown in Fig. 4). Note that (13) allows for dissolution of paint if the surface current density is too small. The workpiece is an equipotential, so

$$\Phi = 0 \tag{14}$$

there; however, this applies on bare patches only as long as

$$-\frac{\partial \Phi}{\partial n} < \delta. \tag{15}$$

If this condition is not satisfied (as, for example, at $t = 0$, when the whole workpiece is bare), (14) is immediately replaced by (12). Lastly, conditions such as $\Phi = 1$ on the anodes are assumed on the exterior boundary of the tank.

As it stands, we have an elaborated version of the Muskat problem. However, we simplify using the fact that $\varepsilon \ll 1$. This means that the paint layer has dimensionless thickness $O(\varepsilon)$, $\varepsilon h(\mathbf{x}, t)$, say, and within it, to lowest order $\Phi$ varies linearly in the direction normal to the workpiece. Thus on $\Gamma$, $\Phi = -h \partial \Phi / \partial n + O(\varepsilon)$, (which is just Ohm's law for the paint layer), and linearising onto the workpiece gives the codimension-two model for $\phi$, the lowest order approximation to $\Phi$ in the codimension two version, as

$$\Delta \phi = 0 \tag{16}$$

in region I', with

$$\phi = 0 \tag{17}$$

on bare patches where $-\partial \phi / \partial n < \delta$; on paint layers $\Gamma'$,

$$-\frac{\partial h}{\partial t} = \frac{\partial \phi}{\partial n} + \delta \tag{18}$$

$$-h\frac{\partial \phi}{\partial n} = \phi \tag{19}$$

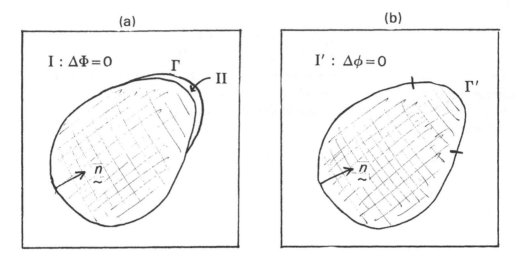

Figure 4: Electropainting: (a) the full problem, (b) the codimension-two version

provided that $h < 0$ and, if $h = 0$, $-\partial\phi/\partial n - \delta > 0$. The free boundary is now at the intersection of bare and painted sections, and there

$$\mid \nabla\phi \mid < \infty. \tag{20}$$

Fixed boundary conditions remain as before.

Some analysis of this problem has been carried out by [Caff], who show that, with suitable anode conditions, the painted areas always expand. This avoids the potential difficulty of dissolution, for in this case the corresponding Muskat problem would appear to be ill-posed, and it is not clear (i) whether the full electropainting problem has a solution at all, and (ii) whether its behaviour is reflected in that of the codimension-two version. We shall return to this point later.

It is of considerable practical interest to be able to calculate the degree to which the inside of a long thin body (for example, a box girder) can be painted. Consider the situation in Fig. 5. It can be tackled by an asymptotic approach similar to that in the previous section. First, note that the interior can only be painted if $\delta$ is small (which is true in practice), for if $\delta$ is large paint will only be deposited near the entrance $x = 0$. In fact the interesting case, in which partial painting can occur, is when the width of the interior is $O(\delta)$, say between $y = \pm a\delta$. Then a nonlinear system of p.d.e.s for the thickness of paint can be derived as follows.

Write $y = \delta Y$ and expand $\phi \sim \phi_0 + \delta^2\phi_1 + \dots$ . It is also necessary to rescale time by $t = \delta^{-1}\tau$; it is a slow process to paint the interior of a box. Just as before,

Figure 5: Electropainting the inside of a box (in two dimensions)

$\phi_0 = \phi_0(x, \tau)$, $\phi_1 = -\frac{1}{2}Y^2 \partial^2 \phi_0 / \partial x^2 + a_1(x, \tau)$. Now applying (18) and (19) on $Y = \pm \alpha$ gives the third-order system

$$-\frac{\partial h}{\partial \tau} = -\alpha \frac{\partial^2 \phi_0}{\partial x^2} + 1, \tag{21}$$

$$\alpha h \frac{\partial^2 \phi_0}{\partial x^2} = \phi_0 \tag{22}$$

for $0 < x < s(\tau)$, where there is paint, and

$$\phi_0 = h = 0 \tag{23}$$

for $x > s(\tau)$. At $x = s(\tau)$, $\phi_0, h$ and (probably) $\partial \phi_0 / \partial x$ are continuous. Also $V_0(t) = \phi_0(0, t)$ and $h(x, 0)$ are prescribed. Apart from the obvious steady solutions and some similarity solutions [Stud], little is known about this system; even the weak formulation does not seem easy, especially if $V_0(t)$ can decrease (although it must remain nonnegative), for then dissolution may occur, with the possibility of new free boundaries.

## 4    Discussion

We have described just two codimension-two boundary problems, from many possible examples. A list of others is given by [Ock], and here we just mention three more.

(i) The elastic contact problem (as usually formulated in Lagrangian coordinates) is already in codimension-two form, and it would only be necessary to perform a linearization as above if it were originally posed in Eulerian variables;

(ii) 'Childress' flow of an inviscid fluid over a small step, with a long thin recirculating flow with constant vorticity [O'Ma];

(iii) Impact of a blunt body on water (see [How] and references therein) , which can also be treated by using Lagrangian coordinates.

We have highlighted the need for a general theoretical approach, for although some of our problems can be treated using variational inequalities and related methods, this is not always true. So far we are not aware of any case in which the full problem is well-posed and the behaviour of the codimension-two version turns out to be markedly different from that of the full problem (taking into account the limitations of the approximation). Furthermore, this property appears to persist even when a 'long thin' limit of the codimension two problem is taken.

Conversely, when the full problem is ill-posed, what evidence there is suggests that so is the codimension-two problem. Perhaps the best-studied case from this point of view is the water entry problem, which appears to be well-posed together with its codimension-two version, while the water exit problem (i.e. remove a blunt body from water) does not. One might also conjecture that the electropainting problem is ill-posed if dissolution occurs, on the grounds that the Muskat problem is probably ill-posed in the corresponding situation. This would then lead to the question of regularisation to ill-posed codimension-two free boundary problems, and their relationship to (corresponding?) regularisations of the full versions – an almost untouched subject.

## Acknowledgements

I should like to thank Anvar Meirmanov, John Morgan, Maura Ughi and especially John Ockendon for many helpful conversations, and the Royal Society for financial support via a University Research Fellowship.

# References

[Aitch 1] Aitchison, J.M., Elliott, C.M. and Ockendon, J.R., 'Percolation in gently sloping beaches', IMA J. Appl. Math. **30**, 269-297, 1983.

[Aitch 2] Aitchison, J.M., Lacey, A.A. and Shillor, M., 'A model for an electropaint process', IMA J. Appl. Math. **33**, 17-31, 1984.

[Caff] Caffarelli, L.A. and Friedman, A., 'A nonlinear evolution problem associated with an electropaint process', SIAM J. Math. Anal. **16**, 955-969, 1985.

[Crank] Crank, J. and Gupta, R.S. 'A moving boundary problem arising from the diffusion of oxygen in absorbing tissue', J. Inst. Math. Appl. **10**, 19-33, 1972.

[Ell] Elliott, C.M. and Friedman, A., 'Analysis of a model of percolation in a gently sloping sand bank', SIAM J. Math. Anal. **16**, 941-954,

1985.

[How]    Howison, S.D., Ockendon, J.R. and Wilson, S.K., 'Incompressible water-entry problems at small deadrise angles', J. Fluid Mech. **222**, 218-230, 1991.

[Ock]    Ockendon, J.R., 'A class of moving boundary problems arising in industry', in *Applied and Industrial Mathematics*, ed. R.Spigler, Kluwer 1991.

[O'Ma]    O'Malley, K., Fitt, A.D., Jones, T.V., Ockendon, J.R. and Wilmott, P. 'Models for high Reynolds number flow down a step', J. Fluid Mech. **222**, 139-155, 1991.

[Stud]    Mathematical Study Group Report, Birmingham 1990 (available from the author).

Mathematical Institute
24-29 St Giles'
Oxford OX1 3LB

## J B KELLER
# Phase fronts in reaction-diffusion problems

### 1. Introduction

Reaction–diffusion equations occur in the description of chemical reactions of reactants which are distributed in space. An example of such an equation is

$$u_t(x,t,\varepsilon) = \varepsilon \Delta u - \varepsilon^{-1} f(u), \quad x \in \Omega.$$

Here $u \in R^m$ is the vector of concentrations of the m reactants and $f(u)$ is the scaled vector of reaction rates. The variables have been scaled to emphasize the smallness of the diffusion coefficient and the largeness of the reaction rate, which are indicated by the factors $\varepsilon$ and $\varepsilon^{-1}$ respectively. The domain containing the reactants is $\Omega$ in $R^n$, and on its boundary we require that

$$\partial_n u(x,t,\varepsilon) = 0, \quad x \in \partial\Omega.$$

Initially the concentrations are specified:

$$u(x,0,\varepsilon) = g(x).$$

There is an extensive mathematical theory of equations like (1.1), and much of it is reported in the book of Fife [1]. The solutions of such equations can develop fronts or internal layers across which they change very rapidly, and the motion of such fronts has been studied by many authors. A recent book by Fife [2] describes some of those studies and gives many references to the literature.

Certain new results concerning the formation and motion of fronts have been obtained by Rubinstein, Sternberg and Keller [3, 4]. Those results and related ones formed the basis for a lecture by the author at the First Venice Conference on Applied Mathematics in October, 1989, which is summarized in the Proceedings of that Conference (Keller [5]). They also were the basis for part of the author's lecture at the International Colloquium on Free Boundary Problems: Theory and Applications, at the Centre de Recherches Mathematiques, Université de Montreal in June, 1990. Therefore this summary of that part of the lecture will be quite brief since the details are available in references [3] – [5].

Another part of that lecture concerned the stability of crystals that grow or evaporate by step propagation. The step mechanism of crystal growth was proposed by Burton, Cabrera and Frank [6] in 1951. It leads to a Stefan–like problem for the motion of a step on a crystal surface. This problem has solutions in which a single step moves at constant velocity, and other solutions in which an infinite number of equally spaced parallel steps move at constant velocity. Recently, Ghez, Cohen and Keller [7] have studied the stability of these solutions to perturbations in which the steps remain straight and parallel, improving upon the previous stability studies. The step spacing and the concentration of adsorbed atoms on the terraces between the steps are perturbed, and the growth or decay of these perturbations is determined. An expression for the growth rate in terms of the parameters of the problem is obtained. In one typical case this result shows that growing cyrstals are stable, while evaporating ones are unstable at small evaporation rates but stable at large evaporation rates. This topic will not be discussed further here.

2.  Phase fronts

To study the asymptotic behavior of the solution of $(1.1) - (1.3)$ as $\varepsilon \to 0$, we consider first the initial time interval, within which it is convenient to introduce the time variable $\tau = t/\varepsilon$. Then $u(x, \tau, \varepsilon)$ satisfies $(1.2)$, $(1.3)$ and

2.1 $$u_\tau = - f(u) + \varepsilon^2 \Delta u, \ x \in \Omega.$$

Upon writing $u = u_0 + \varepsilon u_1 + 0(\varepsilon^2)$ we find that $u_0$ satisfies the ordinary differential equation and initial condition

2.2 $$u_{0_\tau} = - f(u_0), \ u_0(x,0) = g(x).$$

If the ordinary differential equation in $(2.2)$ has exactly one globally attracting fixed point $U_1$ then at every $x$, $u_0$ will tend to $U_1$ during the initial interval. Then $u$ will do so also, and diffusion will play a minor role in modifying the rate of approach to $U_1$. If $\partial_n g(x) = 0$ for $x \in \partial \Omega$ then $u_0$ will satisfy the boundary condition $(1.2)$, but $u_1$ will have a boundary layer at $\partial \Omega$ within which it will become adjusted to the boundary condition.

The more interesting case is that in which $(2.2)$ has two attracting fixed points $U_1$ and $U_2$. Then $\Omega$ can be divided into two subdomains $\Omega_1$ and $\Omega_2$. The subdomain $\Omega_j$ consists of the points $x$ for which $g(x)$ is in the basin of attraction of $U_j$, $j = 1,2$. It follows that $u_0(x,\tau) \to U_j$ as $\tau \to \infty$ if $x \in \Omega_j$. The boundary between $U_1$ and $U_2$ is called a front, and $u_0$ tends to two different values on the two sides of the front. Since this limit function is discontinuous across the front, the expansion $u = u_0 + \varepsilon u_1 + O(\varepsilon^2)$ cannot be valid on and near it. This is clear

25

because the Laplacian in (2.1) would become infinite there.

To construct an expansion which is valid at and near the front we introduce a coordinate which measures distance from the front and we stretch this coordinate by the factor $\varepsilon^{-1}$. Since the front will move, we describe it by a function $\varphi(x,t,\eta)$ where $\eta = \varepsilon t$ is a slow time variable. Then the stretched variable is $z = \varepsilon^{-1}\varphi(x, t, \eta)$ and we write $u(x,t,\varepsilon) = u^0(x,z,\tau,t,\eta) + \varepsilon u^1(z,x,\tau,t,\eta) + O(\varepsilon^2)$. From (2.1) we get equations for $u^0$ and $u^1$, and we assume that the solutions of these equations tend to traveling waves as $\tau \longrightarrow \infty$: $u^0 \sim Q(z - c\tau,x,t,\eta)$, $u^1 \sim P(z - c\tau,x,t,\eta)$. The equations for $u^0$ and $u^1$ lead to ordinary differential equations for Q and P.

We seek a solution for Q which tends to $U_1$ as $z \longrightarrow -\infty$ and to $U_2$ as $z \longrightarrow +\infty$. These conditions determine a particular solution Q. They also lead to an equation for $\varphi$:

$$\frac{c - \varphi_t}{|\nabla\varphi|} = c_0.$$

The left side of (2.3) is just the normal velocity of a level set of $\psi(x,t,\eta) = \varphi(x,t,\eta) - ct$ and the right side is a constant determined by the problem for Q. Thus the fronts move along their normals with constant speed $c_0$. When $f(u) = V_u(u)$ where V is a scalar, then $c_0$ is proportional to [V], the jump in V across the front. Thus the front moves toward the side with the larger value of V(u) so that the ultimate value of u is that with the minimum value of V(u).

When $c_0 = 0$, which occurs when $V(U_1) = V(U_2)$ in the case $f = V_u$, the front does not move on the time scale measured by t. However, then the equation for P shows that $\varphi$ satisfies the equation

$$\frac{\varphi_\eta}{|\nabla\varphi|} = \kappa_\varphi.$$

Here $\kappa_\varphi$ is the mean curvature of the level set $\varphi = $ constant. This equation shows that the level set $\varphi = $ constant moves along its normal with velocity $- \kappa_\varphi$ on the $\eta$ time scale, provided that $c_0 = 0$. When $c_0$ is small of order $\varepsilon$, then we find that the normal velocity of the front, or of any other level set of $\varphi$, is $c_0 - \kappa_\varphi$. A less formal derivation of (2.4) was given by Allen and Cahn [8]. Bronsard and Kohn [9] and de Mottoni and Schatzman [10] have proved that (2.4) is correct in the limit $\varepsilon \longrightarrow 0$ under suitable conditions on $f(u)$ and on the initial data.

The boundary condition (1.2) leads to the condition that the front must be normal to the boundary where they meet.

3.   Flow by curvature

The motion of a surface along its normal, with a velocity proportional to its mean curvature, is called "flow by curvature" by geometers. Their results on such flows can be applied to the motion of fronts. Those results are primarily for motions in unbounded domains, and the strongest results are for curves embedded in the plane. For example Gage and Hamilton [11] showed that if the curve is initially closed and convex it will remain closed and convex, its length will decrease, it will remain inside the original curve, and it will shrink to a point in a finite time. Grayson [12] has shown that any simple closed curve will eventually become convex and then shrink to a point in a finite time.

These results apply to curves in a bounded domain $\Omega$ provided the curves never intersect $\partial\Omega$. Rubinstein, Sternberg and Keller [3] considered two smooth curves in $\Omega$ that are normal to $\partial\Omega$ where they intersect it and that evolve according to flow by curvature. They showed that if they do not intersect each other at $t = 0$, then they never intersect each other. To prove this they first proved a Hopf–type corner lemma for parabolic equations.

The only fronts which do not move are straight line segments, so the only possible steady fronts in $\Omega$ are straight line segments normal to $\partial\Omega$ at both ends. By using the preceding comparison theorem, it can be shown that such segments of locally minimal length are stable under flow by curvature. Thus they can arise as the limits of moving fronts as time becomes infinite, and the corresponding steady solution for u is not constant. Instead it consists of one region where u is nearly equal to $U_1$ and another region where U is close to $U_2$. These regions are separated by a thin transition layer which lies along the straight segment.

The case in which $f(u) = 0$ for all u in a d–dimensional manifold M in $R^m$ was treated in [4], with M being an attractor for (2.2). In that case as $\epsilon \longrightarrow 0$ the time–dependent solution of (1.1) tends to a diffusion with values in M. The solution then tends to a harmonic map of $\Omega$ into M as $t \longrightarrow \infty$. When M consists of two disjoint manifolds $M_1$ and $M_2$ then $\Omega$ consists of two subdomains $\Omega_1$ and $\Omega_2$ such that $u_0(x,\tau)$ tends to $M_j$ as $\tau \longrightarrow \infty$ if $x \epsilon \Omega_j$, $j = 1,2$. Then the boundary between $\Omega_1$ and $\Omega_2$ develops into a moving front. Some aspects of the motion of this front were considered in [4], but much more remains to be done.

27

References

[1] Fife, Paul C. (1979) Mathematical Aspects of Reacting and Diffusing Systems, Springer, New York.

[2] Fife, Paul C. (1988) 'Dynamics of internal layers and diffusive interfaces,' CBMS–NSF Regional Conference Series in Applied Mathematics 53, SIAM, Philadelphia.

[3] Rubinstein, Jacob, Sternberg, Peter, and Keller, Joseph B. (1989) 'Fast Reaction, Slow Diffusion, and Curve Shortening,' SIAM J. Appl. Math. 49, 116 – 133.

[4] Rubinstein, Jacob, Sternberg, Peter and Keller, Joseph B. (1989) 'Reaction–Diffusion Processes and Evolution to Harmonic Maps,' SIAM J. Appl. Math. 49, 1722 – 1733.

[5] Keller, J. B., 'Diffusively coupled dynamical systems,' Venice 1989: The State of the Art in Applied and Industrial Mathematics, Kluwer Academic Publishers, The Netherlands, in press.

[6] Burton, W. K., Cabrera, N., and Frank, F.C., Philos. Trans. R. Soc. (London) A243, 299(1951).

[7] Ghez, R. Cohen, H.G., and Keller, J. B., App. Phys. Lett. 56(20), 1977–1979, 14 May 1990.

[8] Allen, S. W., and Cahn, J. W. (1979) 'A microscopic theory for antiphase boundary motion and its application to antiphase domain coarsening,'Acta Metallurgica 27, 1085–1095.

[9] Bronsard, L., and Kohn, R. (1988) preprint.

[10] de Mottoni, P. and Schatzman, M. (1989) 'Évolution géométrique d'interfaces,' Compt. Rend. 309, 453 – 458.

[11] Gage, M., and Hamilton, R. S. (1986) 'The heat equation shrinking convex plane curves,' J. Differential Geometry 23, 69 – 96.

[12] Grayson, M. (1987) 'The heat equation shrinks embedded plane curves to round points,' J. Differential Geometry 26, 285 – 314.

Joseph B. Keller
Departments of Mathematics and Mechanical Engineering
Stanford University
Stanford, CA 94305

P ORTOLEVA
# Free boundaries in geochemical systems

## I. Introduction

Three classes of free boundary problems in geochemical systems are used to illustrate the richness and importance of problems to be addressed. The growth of solid solution crystals or polycrystalline masses is shown to yield interior oscillatory compositional zoning through a surface-attachment autocatalytic feedback. Reaction fronts driven by the infiltration of reactive fluids into a rock (i.e., a multi-component porous medium) can undergo morphological instabilities and lead to temporally oscillatory precipitation on a wide range of length and time scales. Fracture fronts in 100 kilometer-scale "basins" can undergo a temporally oscillatory instability when compaction and other mechanisms elevate fluid pressures beyond the hydrofracture point.

These phenomena involve distinct free boundaries with structures that have interesting differences. Here we set forth these problems, review some results and pose some outstanding questions.

Two distinct types of challenges exist in geochemical free boundary analysis. One set of approaches addresses the use of scaling and matched asymptotics to derive free boundary problems, analysis techniques to prove their rigorous existence and uniqueness and bifurcation and numerical techniques to derive the behavior of the free boundaries and their instabilities. Another type of challenge is for systems wherein a key part of the problem is that free boundaries are created, destroyed or become diffuse. In the latter case, the scaling of the width of the associated transition layer changes continuously as the system evolves and may even lead to stretching of the layer to the point that a free boundary is no longer well defined. We shall illustrate all these issues via concrete examples.

## II. Free Boundary Nurseries, Variable Scaling and Annihilation of Reaction Fronts in Porous Media

Reaction fronts arise in rocks when fluids are injected that are undersaturated with respect to at least one of the minerals in the rock. When reactions are fast or flows are slow these fronts consist of very narrow transition zones across which one or more minerals change (in grain size and volume fraction of rock occupied). Such fronts and the zones between them occur in many geological contexts including redbed copper deposits, uranium roll front deposits, gold deposits, the generation of petroleum reservoirs and traps, infiltration metasomatism and acid stimulation in the context of petroleum engineering.[1,2] We have considered the asymptotic behavior of these fronts, analyzing a number of them as free boundary problems.[1-13]

While analysis of the free boundary problems arising from the asymptotic analysis has given us insights into these phenomena, there is a serious limitation to this approach. It must be admitted that the development of the matched asymptotic analysis is always, to my knowledge, based on a very clear prior understanding of the nature of the solution of the reaction-transport problem. As a result, the problems that have been analyzed involve at most a few minerals, and no or few reactions in the pore fluid. Furthermore, results obtained on one system usually do not give much insight into other systems.

There are two main reasons for this situation. In multi-mineralic problems it is far from clear how many fronts are created, and which fronts are in some sense narrow and hence may be subjected to a matched asymptotic analysis. Furthermore, not all transition layers necessarily have the same scaling for their thickness and there is often a very complex "nursery" zone in the rock from which various fronts are born and leave.

Models of reactive flows in porous media are generally similar to the following. Let $R_i$ be the radius of a mineral i grain in a system along the x-axis at time t. Then the grain growth/dissolution law is written

$$\frac{\partial R_i}{\partial t} = G_i(\underline{c}) \tag{II.1}$$

where $\underline{c}$ $(= \{c_1, c_2, \ldots c_N\})$ is a set of N pore fluid species concentrations. The $c_\alpha$ $(\alpha = 1, 2, \ldots N)$ satisfy

$$\frac{\partial \phi c_\alpha}{\partial t} = -\frac{\partial}{\partial x} J_\alpha + F_\alpha + \sum_{i=1}^{M} \nu_{\alpha i} 4\pi \rho_i n_i R_i^2 G_i \tag{II.2}$$

for net rates $F_\alpha(\underline{c})$ of reactions in the pore fluid, porosity $\phi$, and stoichiometric coefficients $\nu_{\alpha i}$ in the M mineral system ($i = 1, 2, \ldots M$). The molar density $\rho_i$ is assumed constant and the number of grains of mineral $i$ per rock volume, $n_i$, is assumed here to be independent of time. The flux $J_\alpha$ is usually taken in the form

$$J_\alpha = - \phi D_\alpha \frac{\partial c_\alpha}{\partial x} + \phi v c_\alpha, \qquad (II.3)$$

$D_\alpha$ being the diffusion/dispersion coefficient that can depend on the fluid velocity $v$ and the "texture" $\{n_1, R_1, \ldots n_N, R_N\}$. Note that the porosity $\phi$ can be expressed in terms of the texture:

$$\phi + \sum_{i=1}^{M} \frac{4}{3}\pi R_i^3 n_i = 1. \qquad (II.4)$$

The fluid velocity is usually determined from the conservation of mass conditions for the majority (the "solvent") species (say $\alpha=1$) and an "equation of state" relating $c_1$ to fluid temperature, pressure and $c_2, c_3, \ldots c_N$. In the simplest case $c_{\alpha \neq 1} \ll c_1$, this usually takes the form

$$\frac{\partial}{\partial x}(\phi v) = 0 \qquad (II.5)$$

when the pore fluid is incompressible. The theory is completed by giving the phenomenological laws yielding the dependance of $D_\alpha$ on the texture and $v$ and the reaction rates $F_\alpha$ and $G_i$ on $\underline{c}$.[9]

Free boundaries emerge when the mineral reaction rate coefficient becomes large. In general one may write

$$G_i = k_i E_i(\underline{c}) \tag{II.6}$$

where $k_i$ is the rate coefficient and $E_i(\underline{c})$ is a factor which vanishes when the fluid concentrations are in equilibrium with mineral i. Free boundaries emerge as one or more of the $k_i$ become large.

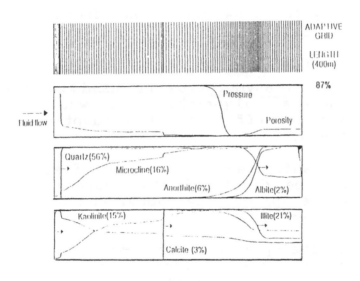

Figure 1

As an illustrative example, consider a numerical simulation of Fig. 1 for an initially uniform, seven mineral system subject to an influx of water for 100,000 years.[14] A number of distinct fronts are emitted from the left (inlet) end and advance downstream to the right. Their speeds are much slower than that (v) of the fluid. Note some fronts are broad and others are narrow. Across each front one or more mineral's volume fraction $\frac{4}{3}\pi R_i^2 n_i$ changes. Note also the adapted grid used to capture the narrow fronts. The various grid spacings suggest rather different scaling

33

of the fronts. Interestingly, these front widths change very distinctly with even small changes in the composition of the inlet fluid because of the highly nonlinear nature of the phenomenological laws. Furthermore, as rate coefficients can depend strongly on temperature, and temperature in a geological system may change along the flow direction, front widths may change as the fronts advance downstream.

From this example we pose the following challenges:

* Can matched asymptotics be used to develop a very general algorithm that would be more accurate and efficient than adaptive gridding techniques;

* Can a sufficiently general free boundary analysis be developed that automatically can create the appropriate number and types of free boundaries;

* Can we develop a variable scaling analysis that bridges a range of front width scaling from very narrow to broad; and

* For the case of a sequence of steady fronts in the multi-mineral system, is the ordering and nature of the sequence of fronts unique for a given inlet fluid and initial rock composition or are there several stable sequences that may be established with different initial data?

Infiltration driven reaction fronts also may display morphological instabilities [3-6,9] (see also the companion paper by Chen and Ortoleva, this volume). For multi-mineralic systems the fingering, branching-tree and other nonlinear spatio-temporal structure of the reaction fronts can be very complex.

## III. Self-Organization at Reaction Fronts Through Ripening and Nucleation Instabilities

The mixing of reactants can lead to the precipitation of solids. When the two reactants are initially separated and then allowed to mix then precipitation could occur in the interdiffusion zone. Interestingly the distribution of this precipitation in space can take on banded or other forms not imposed by the mixing geometry. These patterns are called Liesegang bands.[15]

Liesegang phenomena apparently occur in a variety of geochemical systems in association with reaction fronts. Typically one or more reactants are imported to the front in the imposed fluid and the remaining solutes come from the dissolution of one of the minerals at the front. Consider a system involving pore fluid solutes X and Y and minerals A and B and the processes

$$X + A = Y \tag{III.1}$$

$$X + Y = B. \tag{III.2}$$

Let the medium contain only mineral A. Then an A-dissolution front will advance into the system if the imposed fluid is X-rich and Y-poor. Under this situation a region of B-precipitation can exist upstream of the A dissolution front from the interaction of X imported in the fluid and Y diffusing upstream – see Fig. 2. Note the existence of the A-front, B-pulse complex.

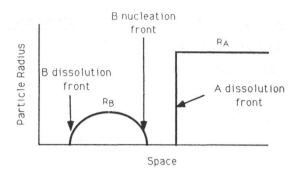

Figure 2

In this phenomena there can be three distinct free boundaries when the rate coefficient is large.

(1) The A-dissolution front

(2) The B-nucleation point

(3) The B-dissolution point

Downstream of the B-nucleation point no B has precipitated while B has been dissolved upstream of the B-dissolution front. An analytical solution of the steady-state front-pulse complex free boundary problem has been obtained.[8]

This type of pulse-front complex system provides us with an opportunity to investigate Liesegang phenomena in terms of a bifurcation analysis. In a classical Liesegang experiment, B precipitation is created by diffusing the pore fluid solute species X and Y into each other in a cross flux configuration. Because of the diffusion there is no steady state and banding is a transient phenomena. In the present geochemical problem there is a steady front-pulse complex solution (i.e., a configuration that is independent of time in a reference frame moving with the complex). Thus we might expect that bands could emerge from the nucleation point and disappear into the dissolution point in a way that is periodic in the complex-fixed frame. The question arises as to the nature of the onset of oscillation – does it arise as a finite frequency-small amplitude (Hopf-type) bifurcation or as a zero-frequency, finite-amplitude transition – see Ref. 16.

Two classes of models of these phenomena have been considered. Nucleation threshold models may be written in terms of the equations of the previous section except that the rate of grain growth is augmented as follows. For the reaction (III.2) the rate $G_B$ is written

$$G_B = k_B(XY - Q) \qquad (III.3)$$

where X and Y are concentrations (in moles/pore fluid volume) and Q is a constant (the "equilibrium constant"). The nucleation of B occurs when the product XY exceeds a "nucleation threshold" $Q_n > Q$. To account for this we write [1,2,8,17]

$$k_B = \begin{cases} 0, \ R_B = 0, \ XY < Q_n \\ k_B^O, \text{ otherwise} \end{cases} \qquad (III.4)$$

for constant $k_B^O$. This model has been solved numerically for a variety of oscillatory cases.[2,8] As system parameters change the B-pulse shakes, gets internal structure (maxima and minima) and then may break up into several pulses to form distinct precipitation banding (see Fig. 3). This hierarchy of transitions to states of high complexity is a rich area for future research.

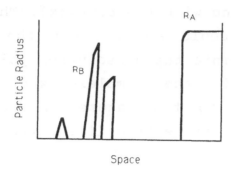

Figure 3

A second class of models make more realistic account of the nucleation process by considering the distribution of particle sizes and surface tension.[1,18] In the context of the A,B model cited above one introduces the particle size distribution $F(R,x,t)$ such that

$$F(R,x,t)dR = \text{the vicinity of x at time t with}$$

number/unit volume of B particles in the vicinity of x at time t with radius in an interval dR about R.

From Lifshitz-Slyouzov (LS) theory [19] F satisfies

$$\frac{\partial F}{\partial t} + \frac{\partial}{\partial R}(G_B F) = 0. \tag{III.5}$$

A key part of the LS model is the dependence of Q of (III.3): $Q(R)$ starts at $Q(\infty)$, rises to a maximum at a critical radius $R_C$ and then decreases back to $Q(\infty)$ as $R \to \infty$. For the present problem F vanishes initially and far in advance of the A-dissolution front. If $Y = 0$ in the inlet the F also vanishes upstream of the complex. Also F vanishes as $R \to \infty$ and satisfies a boundary condition at small R, the precise form of which depends on some detailed technical assumptions of the physical chemistry (see Refs. 1,18,19). The simplest case is that $F$ is defined on $R^* \leq R < \infty$ and $F(R^*,x,t)$ is proportional to $X(x,t)Y(x,t)$. Finally the conservation of mass equations for $X$ and $Y$ take the form (II.2) except that the term $R_B^2 G_B$ is replaced by a new term proportional to

$$\int_{R^*}^{\infty} dR \ R^2 GF. \tag{III.6}$$

For the problem of banding above, $F$ as a function of $R$ has finite support initially; hence it has finite support always and the integral in (III.6) is finite. For this system one must address integro-differential equations in solving the free boundary problem. It remains to analyze the steady front-pulse complex and investigate the dynamic states related to precipitation banding.

## IV. Self-Organization of Zoning and Shape During Crystal, Geode, Agate and Concretion Growth

### A. Background

Single crystals or polycrystalline solid masses commonly grow in geochemical systems and during this growth they take on oscillatory or sector internal compositional zoning or complex morphologies. These patterns are often the result of feedback contained within the network of reaction, transport and mechanical processes and are not the direct reflection of imposed variations in the concretions of the medium in which the inclusion is growing. In this way they are self-organized.[1,17]

Specific examples are abundant. The internal distribution of composition within single crystal, solid solutions may be oscillatory – i.e., there are a number of concentric shells that are alternatingly rich in one or another element. Such intracrystalline compositional zoning may take on more complex forms sometimes known as sector zoning. Sector zoning is often strongly controlled by crystallographic axes. These phenomena occur in crystals grown from magmas, hot aqueous fluids rising from depth or aqueous fluids at near-surface conditions (see Ref. 1 for citations).

Polycrystalline masses may also self organize during growth. Likely examples are agates, geodes, and concretions. These objects may grow within a rock matrix or in a medium (such as sea floor mud) that does not exert stresses on the growing body (see Ref. 1 for citations).

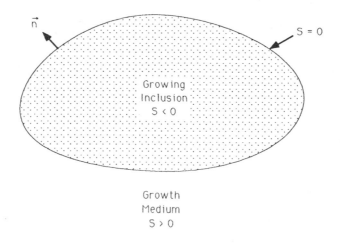

Figure 4

A general case is considered in Fig. 4. The body enlarges from reactants in the environment that typically reach the surface via diffusion. The normal velocity u to the growing body is affected by the composition in the environment at the body surface $(S(\vec{r},t)=0)$; u may also depend on the curvature and on stress when the body is growing in an environment that is highly viscous or even elastic. Thus the body may undergo Mullins-Sekerke morphological instabilities.[20-24] In the case of stress mediated processes there is the extra stabilizing effect of the opposing medium (see Refs. 1 and 25 for further discussion on stress mediated problems).

## B. Catastrophe and Zoning

Here we introduce a model of oscillatory growth based on the chemical kinetics of the addition of formula units to the inclusion surface, and the diffusion in the environment of ions participating in inclusion growth. We focus on the chemistry of the attachment process, assuming that the system is sufficiently undercooled that heat of crystallization effects can be neglected.

## 1. Formulation

The system is described by the concentrations $\underline{c}=\{c_1,c_2,\ldots,c_N\}$ of the N species in the environment and the mole fractions $\underline{f}=\{f_1,f_2,\ldots,f_M\}$ of the M components of the inclusion. Let the inclusion be defined by a function $S(\vec{r},t)$ as in Fig. 4 ($S < 0$ inside the inclusion, $S > 0$ in the environment). From kinematic considerations S evolves via

$$\frac{\partial S}{\partial t} + u|\vec{\nabla}S| = 0 \quad \text{at } S = 0 \qquad\qquad (IV.1)$$

where u is the velocity of inclusion growth normal to the surface; u depends on $\underline{c}$ in the environment near the inclusion surface and $\underline{f}$ in the inclusion near the interface.

In the environment, the participating species can take part in reactions and migrate. If $W_\alpha$ is the net rate of the homogeneous reactions in which species $\alpha$ participates, and $\vec{J}_\alpha$ is its flux then we have the reaction-transport problem in the growth medium:

$$\frac{\partial \phi c_\alpha}{\partial t} = -\vec{\nabla}\cdot\vec{J}_\alpha + W_\alpha , \quad S > 0 . \qquad\qquad (IV.2)$$

To completely describe $\underline{c}$, we must obtain the boundary condition on it at the inclusion surface. Let $\vec{n}$ be an outward normal to the inclusion surface (see Fig. 4). From analytical geometry we have

$$\vec{n} = \vec{\nabla}S/|\vec{\nabla}S| \text{ x.} \qquad\qquad (IV.3)$$

Consider a reference frame attached to the inclusion surface at a given point $\vec{r}$ and time t. The net normal flux away from the surface in this moving frame for $\alpha$ is $\vec{n}\cdot\vec{J}_\alpha - \phi u c_\alpha$, the minus sign arising from the fact that in the reference frame fixed to $S = 0$, the environment appears to move in the $-\vec{n}$ direction. Then if $P_\alpha$ is the net rate of incorporation of $\alpha$ into the inclusion at its surface (moles/area-time), we have

$$\phi u c_\alpha - \vec{n}\cdot\vec{J}_\alpha = P_\alpha . \qquad\qquad (IV.4)$$

Let us assume for simplicity that there is only one reaction creating each type of component on the inclusion surface. If $Y_j$ $(j=1,2,\ldots,M)$ is a formula unit of component j, we write

$$Y_j = \sum_{\alpha=1}^{N} \nu_{\alpha j} X_\alpha \qquad (IV.5)$$

where $X_\alpha$ represents a mobile environment species number $\alpha$ and $\nu_{\alpha j}$ is a stoichiometric coefficient. If $g_j$ represents the rate of reaction (IV.5) (in moles/area-time), we have

$$P_\alpha = \sum_{j=1}^{M} \nu_{\alpha j} g_j \quad . \qquad (IV.6)$$

To complete the theory we must obtain a connection between u and the g's. The number of j-formula units added per unit area in a small time $\delta t$ is, by definition, $g_j \delta t$. In a time $\delta t$, $u\delta t$ is the volume added to the inclusion per unit area. If $\rho(\underline{f})$ is the number of moles of formula units of any type per inclusion volume, then $\rho u\delta t$ is the number of formula units of all types added per unit area in a time $\delta t$. Hence, we have

$$\rho(\underline{f})u = \sum_{j=1}^{M} g_j(\underline{c},\underline{f}) \quad . \qquad (IV.7)$$

Because the LHS of (IV.7) is the total rate of deposition of formula units (moles/area-time) and $g_j$ is the rate of j formula units deposited, then

$$f_j = g_j/\rho u \ , \quad S = 0^- \qquad (IV.8)$$

where $0^-$ is a negative infinitesimal indicating a position just inside the inclusion. With this one may develop the spatial distribution of $\underline{f}(\vec{r})$ in $S(\vec{r},t)$

< 0 as the inclusion grows (see later for further details). This completes the present formal reaction-transport theory of crystal zoning when the form of the incorporation rates $g_j$, the molar density $\rho$ and the transport laws $\vec{J}_\alpha$ are given.

## 2. Catastrophe and Oscillatory Zoning

The preceding model of crystal zoning and growth may be used to demonstrate a mechanism of oscillatory zoning via a surface composition-mediated feedback. Consider the case of plagioclase feldspar crystals growing from a melt.[26] The key point is that the component mole fractions of incorporated formula units at the inclusion surface can flip between albite- and anorthite-rich growth modes. To see this possibility quite generally, we combine (IV.7,8) to obtain the "nonequilibrium fractionation relations" yielding the M mole fractions being incorporated at the surface:

$$g_j(\underline{c},\underline{f}) = f_j \sum_{i=1}^{M} g_i(\underline{c},\underline{f}) , \quad j = 1, 2, ..., M - 1$$

(IV.9)

$$f_1 + f_2 + ..., f_M = 1 .$$

(IV.10)

This constitutes M equations to determine the M mole fractions being incorporated as functions of the environment species concentrations $\underline{c}$ in the immediate vicinity of the crystal/melt interface. An interesting consequence of the present development is that the solution for $\underline{f}$ as a function of $\underline{c}$ via the nonequilibrium fractionation relations (IV.9,10) is not necessarily single valued. Consider for simplicity the binary case, M = 2. Then there is only one independent mole fraction. Letting $f_1 \equiv f$, (IV.9) becomes

$$f\left[g_1(\underline{c},f) + g_2(\underline{c},f)\right] = g_1(\underline{c},f) \,. \qquad (IV.11)$$

To investigate the range of possible zoning phenomena in binary systems within the present model, we must determine the topology of the surfaces in the $N + 1$ dimensional space $(f, c_1, c_2, \ldots, c_N)$ generated by the solution of (IV.11). Catastrophe theory [26-29] indicates that the topological features of such a surface can take the form of the "cuspoids". Catastrophe theory has been used to analyze other problems in self-organizing reaction-transport systems – see Refs. 16,28 and citations therein.

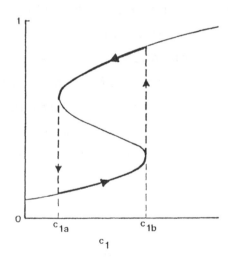

Figure 5

The simplest cuspoid is the fold. It can appear even if only one environment species varies – the others being held constant by either their relative abundance or their relatively large diffusion coefficients. An example of a fold-induced oscillatory zoning dynamics is seen in Fig. 5, where $c_1$ is taken to be the one concentration that is variable. Suppose the system starts on the lower branch at point O. Then $f_2 = 1 - f$ is relatively large. Assume that component 2-rich growth implies that environment species 1 is not consumed. Then $c_1$ at the environment-solid interface will build up. Eventually, $c_1$ may exceed a value $c_{1b}$ such that the lower branch no longer exists (see Fig. 5). Then the inclusion is forced into a component 1-rich growth mode - the upper branch of the nonequilibrium fractionation curve. But in the 1-rich growth mode, environment species 1 is assumed to be consumed and hence $c_1$ decreases; $c_1$ may decrease sufficiently that the left knee of the fractionation curve (at $c_{1a}$) is passed and component 2-rich growth - the lower branch - commences, completing the cycle. Thus, the multiplicity of a nonequilibrium fractionation curve can cause "relaxation oscillation" such that the composition oscillates with rapid switching between growth modes as seen in Fig. 6.

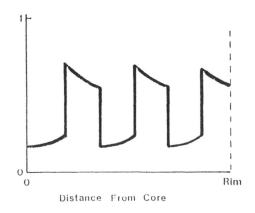

Distance From Core

Figure 6

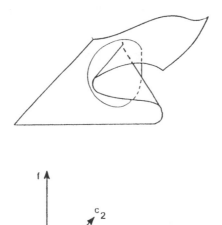

Figure 7

A new level of complexity arises when two environment species may vary. This allows for a "cusp" in the fractionation surface as suggested in Fig. 7. Arguing as earlier, one can envision scenarios where component 2-rich growth would yield an arc in the $c_1,c_2$ plane that could bring the surface mole fraction f to the knee of the lower portion of the fractionation cusp; this could then lead to a switching to 1-rich growth by taking $c_1,c_2$ beyond the lower knee of the fractionation cusp. If the subsequent growth dynamics went around the cusp point, the return to 2-rich growth could be smooth as suggested in Fig. 8. Thus, when there is a cusp in the fractionation surface, oscillatory zoning may proceed with a rapid switching in composition followed by a smooth return, as illustrated in Fig. 8. A number of nonperiodic but undulatory zoning profiles can be attained in association with cuspoid fractionation.

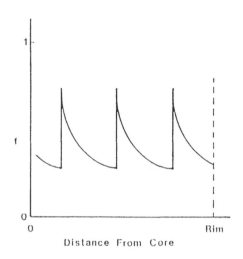

Figure 8

More complex oscillatory f profiles are possible with three or more active environment species. The topological features arising in these more complex binary solid systems are the cuspoids of higher order.[27,28,30] When the inclusion consists of three or more components, there may be even more exotic topological features. Some of these have been identified.[27,30]

The preceding zoning scenarios are all autonomous phenomena. They demonstrate that periodic and more complex undulatory zoning patterns need not arise from undulatory conditions imposed on the environment. Thus, these undulatory phenomena are true examples of self-organization.

## C.  Smooth And Multiple Spatial Dimensional Zoning Patterns

The compositional zoning within an inclusion can take on complex, nonplanar patterns. Furthermore, zoning may not involve discontinuous variations as implied by the formalism of the previous section. We now formulate the inclusion growth problem so as to allow for smooth zoning profiles. A more detailed account of the surface and nonplanar growth involved in more complex pattern spatial geometries will be developed. To avoid notational complexity, the formalism is presented via a simple chemical model.

To resolve the discontinuous zoning behavior of the model of the previous section, a more detailed examination of the surface structure and probabilistic nature of crystal growth has been carried out.[1,31] The result is the following equations for the evolution of the compositional zoning profile.

$$\rho\Delta \frac{\partial f}{\partial t} = -\rho u f + g_A \, , \qquad (IV.12)$$

an equation for the mole fraction f. The discontinuous zoning picture of the previous section emerges directly from (IV.12) when the characteristic length parameter $\Delta$ becomes small. If $\Delta$ is small, the LHS of (IV.12) vanishes and we regain the nonequilibrium fractionation relation introduced in Section B. For polycrystalline masses $\Delta$ is on the order of the local average crystal size.

The statistically averaged nonequilibrium fractionation relation (IV.12) can be written more explicitly. Let the inclusion be contained in the volume $S(\vec{r},t) < 0$ defined by the inclusion surface $S(\vec{r},t) = 0$. Then (IV.12) takes the form

$$u\rho\Delta\vec{\nabla}S\cdot\vec{\nabla}f = \{-\rho uf + G_A\}|\vec{\nabla}S| \ . \tag{IV.13}$$

Thus, the time derivative of $f(\vec{r})$ constrained to points $\vec{r}$ moving with the inclusion surface is $u\vec{n} \cdot \vec{\nabla}f$.

Let us examine how (IV.12) operates in more detail. Consider growth that does not deviate too far from planarity. The inclusion-host interface function S can be taken in the form $S = z - Z(x,y,t)$, where Z is the value of the z-coordinate (assumed unique) for a given position in the x,y plane upon which the interface can be projected. With this the interface dynamics equation (IV.1) reads

$$\rho \frac{\partial Z}{\partial t} = (g_A + g_B)\left[1 + z_x^2 + z_y^2\right]^{1/2} \tag{IV.14}$$

where $z_x = \partial Z/\partial x$ and $z_y = \partial Z/\partial y$. Because f at the surface is $f(x,y,Z)$, (IV.12) becomes, for constant $\Delta$,

$$\Delta\left[\frac{\partial f}{\partial z} - z_x\frac{\partial f}{\partial x} - z_y\frac{\partial f}{\partial y}\right](x,y,z)$$

$$= \left\{-f + \frac{g_A}{g_A + g_B}\right\}\left[1 + z_x^2 + z_y^2\right]^{1/2}. \tag{IV.15}$$

From this we see that when f varies smoothly (on a scale much greater than $\Delta$), the LHS of (IV.15) is small and we regain the discontinuous theory. In regions where f varies rapidly, however, the LHS of (IV.15)

50

cannot be neglected and the discontinuous behavior is resolved.

Because the present model allows for completely smooth variations of solid composition, standard linear methods can be used to study the stability of steady-state inclusion growth to oscillatory zoning. Furthermore, a Hopf bifurcation theory of the oscillation could be obtained for conditions near the onset of instability. Two- and three-dimensional patterning phenomena can also be investigated using stability and bifurcation theory methods. Because the formalism is fully three dimensional, we may also investigate possible complex mixed morphological and compositional instabilities, generalizing the work of Mullins and Sekerke and more recent work on the morphological stability of growing monomineralic solids.[20-22]. An investigation of the limiting behavior as $\Delta \to 0$ will allow more rigorous analysis of the discontinuous oscillations and, most interestingly, more complex two- and three-dimensional compositional and morphological patterns as has been done for chemical waves.[16]

V. Oscillatory Instability of Large Space and Time Scale Fracture Fronts in Compartmented Basins

A sedimentary basin is a depression in the earth's crust into which sediment is deposited. The shape of the basin is determined by the distribution of stresses at the boundary of the basin, and the rate and pattern of sediment input and the reaction, transport and mechanical (RTM) processes within the basin. As a result, this 100 kilometer-scale domain changes shape on the 100 million year time scale, develops moving domains of fracturing and rock cementation, petroleum is produced, migrates and is trapped and sediment is transformed into rock. The sedimentary basin is rich in free boundaries, many of which are of great economic importance.

It is beyond the scope of the present paper to give a review of the many free boundaries we have identified in our analysis and numerical simulations of the RTM dynamics of a sedimentary basin. For illustrative purposes, consider the problem of hydrofracturing and related fronts, oscillatory fronts and auto-isolation effects leading to basin compartmentation.[32-34]

Hydrofracturing occurs when the pore fluid pressure exceeds the least principal stress. The most common situation is that the vertical stress (due to the weight of the overlying rock) exceeds the horizontal stress. Thus if fluid pressures at depth get sufficiently large, fractures open to relieve it and the plane of the fracture is generally in the vertical direction.

There are three major mechanisms of developing pressures at depth that exceed the hydrostatic fluid pressure:

*   thermal expansion of the fluid;
*   compaction of the rock matrix; and
*   chemical reactions (such as petroleum production) for increased volume.

It is the overpressure (amount by which pressure exceeds hydrostatic pressure) that drives fluid flow.

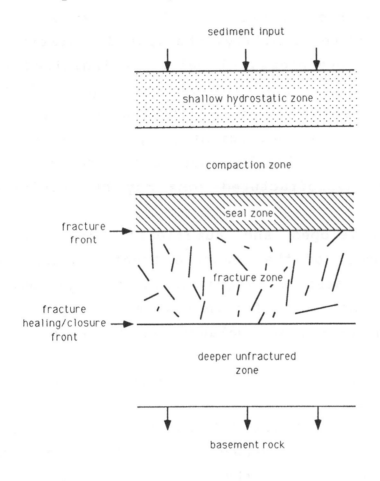

Figure 9

As a result of the interplay of hydrofracturing, compaction and fluid flow, a fracture front may advance upward through the incoming sediment pile as suggested in Fig. 9. In the shallow parts of the basin fluid pressures are typically hydrostatic (except as driven by boundary effects at the surface). Deeper rock is

compacted and hence porosity and permeability decrease
with depth there; fluid pressures do not differ
appreciably from hydrostatic here also. Because of the
region of reduced permeability, the upward migration of
fluids that are overpressured below is inhibited and
hence significant overpressure can develop at depth.
Thus there may be a zone of hydrofracturing below the
compaction zone. At sufficiently great depth the
horizontal stress typically approaches the vertical
stress and the hydrofractured zone may be terminated
from below.

We have developed and numerically simulated a
mathematical model of the fracture front problem. The
model is complicated as required by the geological
system. Aspects accounted for include macroscopic force
balance, grain-scale force balance, chemical reaction-
mediated compaction, fluid flow (via Darcy's law),
water and solute mass balance, fracturing and poro-
elastic rock rheology.

Our simulations of the RTM basin model show that
the transition zone between the fractured and
unfractured zones can be rather narrow. This is because
the kinetics of fracture growth is very rapid once
fracturing commences. This suggests that an asymptotic
analysis of the model will yield a free boundary across
which the fracture size and the rock permeability and
porosity will jump.

The fracture front can apparently undergo a Hopf
bifurcation as suggested in Fig. 10. Assume that the
sediment input rate and other parameters are constant

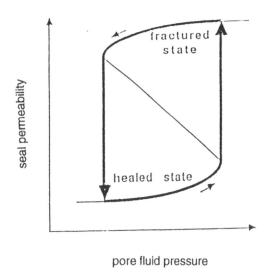

Figure 10

so that there can be a steady state fracture front. If we perturb the front upward then it extends into regions where permeability is relatively large – higher than the permeability minimum or "sealing" zone lying on top of the fracture front. As a consequence fluid can escape upward from the fractured zone and the pressure there drops. Fractures will then heal and the front will move down lower than its steady state position; but the pressurizing mechanisms will eventually raise the pressure in the lower part of the fractured zone because the minimum of permeability has now dropped, as has its value, because pressure had been lower there. Driven by the developed overpressure, the fracture front can now rise above its steady level, completing the cycle. This cycle is illustrated in the simulation of Fig. 11.

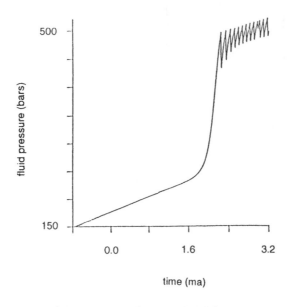

Figure 11

The problem gets even more interesting in two and three dimensions. Observations show that a sedimentary basin may contain "compartments". These are defined as regions of high hydrologic connectivity surrounded by a shell of very low permeability "sealing" rock.[34-36] The mechanisms of generating these seals that comprise a three dimensional isolation of the interior zone is an area of active research at present.[32] Results of geologic data analysis and of simulations of RTM models indicate that there are a number of compartment types distinguished by the types of seals that bound them and the pattern of sedimentation, faulting and stress and thermal history of the basin.

56

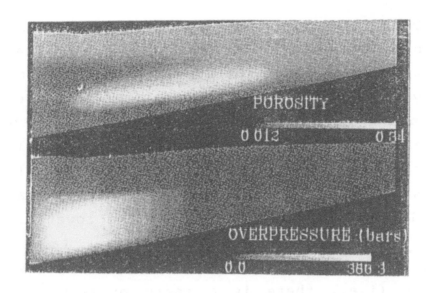

PURUSITY

0.012          0.34

OVERPRESSURE (bars)

0.0          386.3

Figure 12

On a general level, compartmentation is a far-from-equilibrium phenomenon.[1,37,38] If sediment is added very slowly then fluid escape upward could keep up with compaction and the basin would always be close to its equilibrium state. The latter is, among other things, the state where the fluid is on top and porosity-free rock is on the bottom. As the system is driven far-from-equilibrium, some fluid can become trapped in domains surrounded by low permeability rock. An example of a complex compartmentation phenomenon is seen in the RTM simulation of Fig. 12. Here we see the gray level contours of the porosity and the overpressure; the latter preserves the former by repressing compaction and causing hydrofractures in the region of high overpressure. The fracturing front denoting the boundary of the interior of the compartment is seen to take on a characteristic box shape. The shape is influenced by the fault (here modeled as a hydrostatic boundary at the left) and the pattern of sediment input (here in the form of four layers – not shown) but is not a simple reflection of them. Thus, for example, the seal cuts across sedimentary layers. Interestingly, the overpressure in the fractured zone keeps the fractures open and represses compaction while the lower pressure in the seal allows for compaction (and hence permeability loss) and the healing of or repression of fractures (i.e., there is a sharp pressure gradient across the seal).

## Acknowledgements

Research supported in part by a contract with the Gas Research Institute and grants from the U.S. Department of Energy and the National Science Foundation.

## References

1. P. Ortoleva, <u>Geochemical Self-Organization</u>, Oxford University Press, in press (1992).

2. P. Ortoleva, G. Auchmuty, J. Chadam, J. Hettmer, E. Merino, C. Moore and E. Ripley, "Redox front propagation and banding modalities", Physica 19D, 334-354 (1986).

3. J. Chadam, D. Hoff, E. Merino, A. Sen, and P. Ortoleva, "Reactive infiltration instabilities", I.M.A. Journal of Applied Mathematics, 36, 207-221 (1986).

4. J. Chadam, P. Ortoleva and A. Sen, "A weakly nonlinear stability analysis of the reactive infiltration interface", SIAM, Journal of Applied Math, 48, 1362-1378 (1988).

5.  J. Chadam and P. Ortoleva, "A mathematical problem in geochemistry: The reaction-infiltration instability", Rocky Mountain Journal of Applied Mathematics, 21, 631–643 (1990).

6.  W. Chen and P. Ortoleva, "Self-organization in far-from-equilibrium reactive porous media subject to reaction front fingering", in Patterns, Defects and Materials Instabilities, D. Walgraef and N.M. Ghoniem, eds., NATO ASI Series, Vol. 183, 203–220 (1990).

7.  W. Chen and P. Ortoleva, "Development of complex reaction-infiltration front morphologies through nonlinear fluid-rock interaction: Modeling asymptotic and numerical studies", in Proceedings of the SIAM Conference on Mathematical and Computational Issues in Geophysical Fluid and Solid Mechanics, Houston, September 25–29, Frontiers in Applied Mathematics Series, in press (1992).

8.  R. Sultan and P. Ortoleva, "Bifurcation of the Ostwald-Liesegang Supersaturation-Nucleation-Depletion Cycle", in Self-Organization in Geological Systems: Proceedings of a Workshop held 26–30 June 1988, University of California Santa Barbara, P. Ortoleva, B. Hallet, A. McBirney, I. Meshri, R. Reeder and P. Williams, eds., Earth Science Reviews, 29, 163–173 (1990).

9.  W. Chen, A. Ghaith, A. Park and P. Ortoleva, "Diagenesis through coupled processes: Modeling approach, self-organization and implications for exploration", in Prediction of Reservoir Quality Through Chemical Modeling, I. Meshri and P. Ortoleva, eds., AAPG Memoir 49, 103–130 (1990).

10. C.H. Moore and P. Ortoleva, "The effect of fluid and rock compositions on diagenesis: A modeling investigation", in Prediction of Reservoir Quality Through Chemical Modeling, I. Meshri and P. Ortoleva, eds., AAPG Memoir 49, 131–146 (1990).

11. H.K. Haskin, C.H. Moore and P. Ortoleva, "Modeling acid stimulation of the Halfway Formation, Canada, using a geochemical computer model", Society of Petroleum Engineers 18133, 283-294 (1988).

12. P. Ortoleva, G. Auchmuty, J. Chadam, E. Merino and E. Ripley, "Moving redox fronts and the worm hole-cave cascade" in Proceedings of the DOE Argonne Conference on "Nonlinear Problems in Energy Research", (1983).

13. P. Ortoleva, E. Ripley, E. Merino and C. Moore, "Mineral zoning in sediment-hosted copper deposits", in Handbook and Stratiform Ore Deposits, Vol. 13, K.H. Wolf, ed., Amsterdam, Elsevier, 237-260 (1985).

14. J.L. Potdevin and P. Ortoleva, "CIRF: A general reaction-transport code: Mineralization fronts due to the infiltration of reactive fluids", in Proceedings of the 7th Symposium on Water-Rock Interaction, (1992).

15. R.E. Liesegang, Geologische Diffusionen, Dresden, Steinkopff, (1913).

16. P. Ortoleva, Nonlinear Chemical Waves, New York, Wiley, (1992).

17. P. Ortoleva, E. Merino, J. Chadam and C.H. Moore, "Geochemical self-organization I: Reaction-transport feedbacks and modeling approach", American Journal of Science, 287, 979-1007 (1987).

18. R. Lovett, P. Ortoleva and J. Ross, "Kinetic instabilities in first order phase transitions", Journal of Chemical Physics, 69, 947 (1978).

19. I.M. Lifshitz and V.V. Slyouzof, "The kinetics of precipitation from supersaturated solid solutions", Journal of Physical Chemistry Solids, 19, 35-50 (1961).

20. W.W. Mullins and R.F. Sekerke, "Morphological stability of a particle growing by diffusion or heat flow", Journal of Applied Physics, 34, 323–329, (1963).

21. W.W. Mullins and R.F. Sekerke, "Stability of a planar interface during solidification of a dilute binary alloy", Journal of Applied Physics, 35, 444–451, (1964).

22. J. Chadam, S. Howison and P. Ortoleva, "Existence and stability for spherical crystals growing in a supersaturated solution", I.M.A. Journal of Applied Mathematics, 39, 1–15, (1987).

23. J. Chadam and P. Ortoleva, "Moving interfaces and their stability: Applications to chemical waves and solidification", review in Nonlinear Phenomena in Chemistry, I. Hlavacek, ed., New York, Gordon and Breach, (1984).

24. J. Chadam and P. Ortoleva, "Generation of evenly spaced pressure-solution seams during (late) diagenesis: A kinetic theory", Contributions to Mineralogy and Petrology, 82, 360–370, (1983).

25. T. Dewers and P. Ortoleva, "Force of crystallization during the growth of siliceous concretions", Geology 18, 204–207, (1990).

26. P. Ortoleva, S. Haase, J. Chadam and D. Feinn, "Oscillatory zoning in plagioclase feldspar", Science, 209, 272, (1980).

27. A.E.R. Woodcock and T. Poston, A Geometrical Study of the Elementary Catastrophes, Lecture Notes in Mathematics, No. 373, Berlin, Springer-Verlag, (1974).

28. D. Feinn and P. Ortoleva, "Catastrophe and propagation in chemical reactions", Journal of Chemical Physics, 67, 2119 (1977).

29. D. Feinn, W. Scalf, S. Schmidt, M. Wolff, and P. Ortoleva, "Spontaneous pattern formation in precipitating systems", Journal of Chemical Physics, 67, 3771, (1978).

30. R. Thom, Stability, Structure and Morphogenesis, New York, Benjamin, (1972).

31. P. Ortoleva, "Role of attachment kinetic feedback in the oscillatory zoning of crystals grown from melts", in Self-Organization in Geological Systems: Proceedings of a Workshop held 26-30 June 1988, University of California Santa Barbara, P. Ortoleva, B. Hallet, A. McBirney, I. Meshri, R. Reeder and P. Williams, eds., Earth Science Reviews, 29, 3-8, (1990).

32. P. Ortoleva and Z. Al-Shaieb, "Genesis and dynamics of basin compartments and seals", American Journal of Science, accepted, (1992).

33. T. Dewers and P. Ortoleva, "Nonlinear dynamical aspects of deep basin hydrology: Fluid compartment formation and episodic fluid release", American Journal of Science, accepted, (1992).

34. D. Powley, "Pressures, hydrogeology and large scale seals in petroleum basins", in Self-Organization in Geological Systems: Proceedings of a Workshop held 26-30 June 1988, University of California Santa Barbara, P. Ortoleva, B. Hallet, A. McBirney, I. Meshri, R. Reeder and P. Williams, eds., Earth Science Reviews, 29, 215-226, (1990).

35. J.S. Bradley, "Abnormal formation pressure", AAPG Bulletin, 59, 957-973, (1975).

36. J.S. Bradley and D. Powley, "Pressure compartments in sedimentary basins: A review", in AAPG Memoir: Deep Basin Compartments and Seals, P. Ortoleva, ed., (1992).

37. P. Ortoleva, "Far-from-equilibrium dynamics of the compartmented basin", in AAPG Memoir: Deep Basin Compartments and Seals, P. Ortoleva, ed., (1992).

38. T. Dewers and P. Ortoleva, "The role of geochemical self-organization in the migration and trapping of hydrocarbons", Applied Geochemistry, 3, 287-316, (1988).

Peter Ortoleva
Department of Chemistry
Indiana University
Bloomington, IN 47405

V V PUKHNACHOV

# Lagrangian coordinates in free boundary problems for parabolic equations

In this paper we shall be considering the method of Lagrangian coordinates applied to the investigation of Stefan and Verigin problems, and Cauchy problem for degenerate parabolic equations. Besides, finding a hidden symmetry of evolution equations and new classes of quasi-linear equations linearized by the transition to Lagrangian coordinates will be described. The review of related articles, published until 1986, is presented [1].

1. The idea of the method under consideration may be explained by the example of a multi-dimensional one-phase Stefan problem for a linear heat equation. The simplest variant of this problem may be formulated as follows: to find the domain $\Omega_t \in \mathcal{R}^n$ with the boundary $\Gamma_t$ and the function $\Theta(x, t)$ from the conditions:

$$\Theta_t = \Delta\Theta, \; x \in \Omega_t, \; t \in (0, T) \tag{1}$$

$$\Theta = 0, \; \Theta_t - |\nabla\Theta|^2 = 0, \; x \in \Gamma_t, \; t \in (0, T) \tag{2}$$

$$\Theta(x, 0) = \Theta_0(x), \; x \in \bar{\Omega}_0 \tag{3}$$

The domain $\Omega_0$ and the function $\Theta_0$ in initial condition (3) have been prescribed. It is assumed that $\Theta_0 > 0$ if $x \in \Omega_0$ and $\Theta_0 = 0$ if $x \in \Gamma_0$.

Let us consider (1) as the equation of continuity of a medium with density $\rho = \Theta + 1$ and velocity vector $v = -(\Theta + 1)^{-1}\nabla\Theta$. The last equality may be treated as an equation of particle trajectory having coordinates $x$ and velocity $v = x_t$ at time $t$. An initial position of a particle is fixed by the equality

$$x = \xi \quad when \quad t = 0 \tag{4}$$

Proceed in (1), (2) to new independent variables $\xi$ and $t$, and the new unknown functions $x$ and $\rho$. The transformed equation (1) admits integration over $t$ that, taking into account (3) and (4), yields $\rho|M| = \rho_0(\xi) \equiv \Theta_0 + 1$. Here $M$ is the Jacobian matrix of the mapping $\xi \to x$. Solving this equation for $\Theta$ and substituting the resulting solution into the equation of trajectory, we obtain the required system of equations for the functions $x_i(\xi, t), i = 1, ..., n$

$$M^* \cdot x_t = \nabla_\xi \ln(\rho_0^{-1}|M|) \tag{5}$$

Relation (4) represents the initial condition for system (5). When defining the boundary conditions, it should be noted that due to (2) the surface of a unit level of the functions $\rho(\xi, t)$ is independent of $t$. Therefore, if $\Gamma_t$ is built as an image of $\Gamma_0$ on the mapping $\xi \to x$, the second condition of (2) is automatically fulfilled. As far as the first condition of (2) is concerned, it, along with the equality $\rho_0 = 1$ for $\xi \in \Gamma_0$ , implies the relation:

$$|M| = 1 \ when \ \xi \in \Gamma_o \tag{6}$$

The Stefan problem may be formulated now in Lagrangian coordinates [2]. For the cylinder $Q_T = \{(\xi, t) : \xi \in \Omega_0, t \in (0, T)\}$ it is required to find the solution $x = (x_1, ..., x_n)$ of the system of equations (5) satisfying condition (4) on the lower base and condition (6) on the lateral surface.

When proceeding to the Lagrangian coordinates, we obtain a system of quasi-linear equations (5) instead of a linear heat equation. However, the solution definition domain becomes therewith fixed. This gives some advantages when obtaining asymptotical expansions of the Stefan problem solution, analyzing its self-similar solutions, and studying the melting front stability.

2. Let $x(\xi, t)$ be a solution of problem (4)-(6). Evolution of small disturbances of this solution may be studied on the basis of a linear model of the above-mentioned problem, formulated as follows [2]; it is required to define the vector-function $Y(\xi, t)$ satisfying in the $Q_T$ system of equations

$$A \cdot Y_t = \nabla(divY + a \cdot Y) + f \tag{7}$$

the condition

$$divY + a \cdot Y = -q(\nu \cdot Y) + \phi \tag{8}$$

on the lateral surface of

$$Y = 0 \ when \ \xi \in \Omega_0, \ t = 0 \tag{9}$$

Here the symbols $\nabla$ and $div$ denote the differential operators in variables $\xi_1, ..., \xi_n$. The vector $Y$ is associated with the disturbance $X$ of the solution $x$ by the equality $Y = M^{-1}X$. The matrix $A$, the vector $a$ and the scalar function $q$ defined on $S_T$ are in the form: $A = M^*M$, $a = \nabla \ln \rho_0$, $q = \rho^{-1}|\nabla \rho|S_T$ where $\rho = \rho_0|M|^{-1}$ (The solution $x$ is assumed to be rather smooth and such that $|M| \neq 0$ in $Q_T$). Functions $f$ and $\phi$ are prescribed; $\nu$ is the unit vector of outer normal to $\Gamma_0$; $S_T = \Gamma_0 \times (0, T)$.

The type of problem (7)-(9) is indefinite, as evidenced by nonclassical character of the initial nonlinear problem (4)-(6). The theorem of existence and uniqueness of the generalized solution to problem (7)-(9) is proved in [2]. The proof is based on the fact that the inequality $q \geq \delta = const > 0$ providing a necessary a priori estimate of the vector-function $Y$ is fulfilled on the solution of problem (4)-(6).

3. Now let us consider a multi-dimensional self-similar Stefan problem. In this case the domain $\Omega_0$ is a cone in $\mathcal{R}^n$ and $\Theta_0(x)$ is a uniform function of zero homogeneity degree. In Lagrangian variables its solution is reduced to finding the vector-function $f(\xi) = (f_1, ..., f_n)$ in some cone $K = \Omega_0$ so that the following equations are fulfilled

$$N^* \cdot (f - \xi \cdot \nabla f) = 2\nabla \ln((f + \Omega_0)^{-1}|N|), \ \xi \in K \tag{10}$$

$$|N| = 1, \ \xi \in \partial K, \ f \to \xi \ when \ \xi \to \infty \tag{11}$$

where $N$ is the Jacobian matrix with elements $N_{ij} = \partial f_i / \partial \xi_{ij}; i, j = 1, ..., n$. If $f(\xi)$ satisfies (10) and (11), the equalities

$$x = \sqrt{t}f(\xi/\sqrt{t}), \ \Theta = (1 + \Theta_0(\xi))|N(\xi)|^{-1} - 1$$

give a parametric representation of the solution to an initial self-similar problem. Assuming in the first of them $\xi \in \partial K$ we obtain the equation of free surface $\Gamma_t$.

Assume $\omega_1 = \{\xi \in K, |\xi| = 1\}$ and denote the contraction of the function $\Theta_0$ to the domain $\omega_1$ by $u(\phi)$ $(\tau, \phi = (\phi_1, ..., \phi_{n-1})$ are the spherical coordinates in space $\xi$). Assume also that $\partial \omega_1 \in C^{[n/2]+1}$, and $u \in H^3(\omega_1)$ and, besides, the domain $\omega_1$ is such that here all the eigenvalues of the first and second boundary-value problems for the Laplace-Beltrami operator are higher than 2n. The existence and uniqueness theorem of problem (10), (11) for rather small values of $\|u\|_{H^3(\omega_1)}$ is proved in [3] taking into account the above-mentioned assumptions. The proof is based on the solution of the linearized problem in special weight classes of functions [4]. Studying the asymptotics of the solution to problem (10), (11) when $\tau \to 0$ makes possible the description of $t$ smoothing a free boundary as $\Gamma_t$ increases initially having a conic point.

4. In [5] the equation $u_t = \Delta(u^m)$ $(m=const.> 1)$ describing the polytropic gas filtration in a porous medium is solved using the Lagrangian coordinates. Besides, a high-accuracy algorithm based on Lagrange formulation to solve numerically a one-dimensional Cauchy problem for the said equation is presented.

In the previous works the Lagrangian coordinates were used to calculate the phenomena of wave beam self-focusing in nonlinear media [6]. In the above-mentioned problem, a complex-valued function of two real variables is the required one, its modulus square satisfying the equation formally coinciding with that of continuity of one-dimensional radial-symmetric motion.

5. The remaining part deals with the evolution equations with one space variable. Let us consider the equation

$$u_t = [\phi(u, u_x)]_x \tag{12}$$

where $\phi(u, p)$ is an arbitrary function. As shown in [7] in (12), substitution of $x$ by a new independent variable $\xi$ and of $u(x, t)$ by a new required function $\omega(\xi, t)$ according to

$$\xi = \int_0^x u(y, t)dy + \int_0^t \phi[u(0, s), u_x(0, s)]ds \qquad (13)$$

$$\omega(\xi, t) = [u(x, t)]^{-1}$$

transforms it into

$$\omega_t = [-\omega\phi(\omega^{-1}, -\omega^{-3}\omega_\xi)]_\xi \qquad (14)$$

According to [8], transformation (13) may be treated as a transition to a mass Lagrangian coordinate in (12) (this fact was revealed in [9] applied to (12) with $\phi = u^{m-1}$, $m > 1$).

Let us call the substitution of variables in (13) the L-transformation. It maintains the conformity of (12) to equation (14) belonging to the same class:

$$\phi(u, p) \rightarrow \psi(u, p) \equiv -u\phi(u^{-1}, -u^{-3}p) \qquad (15)$$

In [7,10] the analogies of the L-transformation were built for higher-order divergent equations and for (12) where the variables $x, t$ or one of them are complex; thereby the function $u$ is assumed analytical with respect to the corresponding variable.

6. The relationship (15) implies the equality $-u\psi(u^{-1}, -u^{-3}p) = \phi(u, p)$ which means that the L-transformation is of involutive type. The equations describing the sub-class of equations (12) transforming into themselves under the action of L-transformations will be called the L-invariant ones. The comparison between (12) and (14) gives the condition of L-invariance:

$$\phi(u, p) = -u\phi(u^{-1}, -u^{-3}p) \qquad (16)$$

which may be considered as a functional equation with respect to the function $\phi$. The classes of partial solutions (16) containing the arbitrary functions of one variable were found in [7, 8, 11]. A general solution of equation (16) for positive values of $u$ is given by the formula

$$\phi = u^{1/2}F(u, u^{-3/2}p)$$

where a continuous function $F(u, q)$ satisfies the relation

$$F(u, q) = -F(u^{-1}, -q)$$

which is arbitrary otherwise. Thus a class of L-invariant equations (12) is rather wide.

Transformation (13) prescribes automorphism in the set of solutions of L-invariant equations. This property may be useful in the study of unbounded solutions of equation (12). If the function $u(x, t)$ satisfies (12) and vanishes on some line $x = \zeta(t)$ (in particular, if $u$ is the x-odd function), the solution defined by the second equality (13) turns to infinity at the points $\xi = S(t)$ of the image of this line on mapping (13). It should also be noted that the third-order L-invariant equations of the form $u_t = [\phi(u, u_x, u_{xx})]_x$ have been obtained in [10]. The Dym equation $u_t = (-2u^{-1/2})_{xxx}$ is the simplest among them.

7. Application of the L-transformation to equation (12) allows the equation of the form (14) to be derived which admits the Lie group wider than that for equation (12). For example, the equation $u_t = [u^{-2} exp(u^{-1} u_x]_x$ admits a three parameter Lie group. Under the action of the L-transformation it provides the equation $\omega_t = [exp(\omega)\omega_\xi]_\xi$ possessing a four parameter Lie group. The group of local symmetries of the equation $u_t = (u^{-2/3} u_x)_x$ has four parameter. As a result of the L-transformation, we come to the equation $\omega_t = (\omega^{-4/3}\omega_\xi)_\xi$ which is extraordinary among nonlinear equations of the form (12): it admits a five parameter Lie group [12].

The property of expansion of a group, admitted by the equation (12), due to nonlocal transformations is natural to be called the hidden symmetry of this equation [10]. In [8, 10] the exact solutions of such equations, which are invariant with respect to nonlocal symmetries, are presented. The authors of [13] have developed the theory of nonlocal symmetries of evolution equations and have applied it to the group analysis of gas dynamics equations (refer to [14] and the literature cited therein).

8. The example of the equation of the form (12) linearized by non-local transformation (13) was likely to be considered for the first time in [15]. This equation describes heat conduction in solid hydrogen [16]:

$$u_t = (u^{-2} u_x)_x \tag{17}$$

Linearization established in [17] is due to nonlocal substitution of variables in the equation of type (17) with a constant heat source,

$$u_t = (u^{-2} u_x)_x + 1$$

and in [18] linearization of the equation

$$u_t = (u^{-2} u_x)_x - u^{-1})_x$$

is considered, which arises in one model of filtration of two immiscible liquids in a porous medium. Other examples of the equations which are transformed into linear ones by non-local transformations of type (13) are presented in [8]:

$$u_t = -[(u_x)^{-1} + h(u)]_x$$

$$u_t = [(u^{-3}u_x)_x + \mu(t)u^{-2}u_x]_x$$

$$u_t = -u^{3/2}(u^{1/2}u_{xxx})_x$$

Here $h(u), \mu(t)$ are the arbitrary functions.

8. The L-transformation enables us to construct exact solutions of some free-boundary problems possessing a functional arbitrariness. Let us consider a one-phase Stefan problem, where it is required to find the function $\zeta(t)$, $t > 0$, and the solution $u(x,t)$ of equation (17) in the domain $-L < x < \zeta(t)$ such that the following conditions are fulfilled:

$$u = 1, \quad u^{-2}u_x = \frac{dS}{dt} \quad \text{when} \ \ x = \zeta(t), \, t \to 0 \tag{18}$$

$$-u^{-2}u_x = q(t) \ \ \text{when} \ \ x = -L, t > o \tag{19}$$

$$u = u_0(x) \ \ \text{when} \ \ -L \leq x \leq 0, \, t = 0, \, S(0) = 0 \tag{20}$$

Here $L$ is the arbitrary positive constant.

The solutions are exact when

$$q = \frac{G^2}{2\sqrt{(G^2t + a^2)}} \tag{21}$$

where $G > 0$ is the arbitrary constant,

$$a = \int_{-L}^{0} u_0(x)dx$$

$u_0$ is the arbitrary function belonging to the class $C^{2+\alpha}[-L,0]$, $\alpha \in (0,1)$ is such that $u_0 > l$ when $-L \leq x < 0, u_0(0) = 1$ and $u_0^2(-L)u_0'(-L) = B^2/2a$. In the Lagrangian variables we obtain a linear heat equation $\omega_t = \omega_{\xi\xi}$ which is to be solved in the predetermined domain. The boundary condition (21) leads to an explicit expression of $\omega(\xi,t)$ in the form of Fourier series over the function of a parabolic cylinder. Having obtained an appropriate solution of (17)-(21) we shall express the function $S(t)$ determining a free boundary as follows:

$$S = C_0(G^2t + a)^{1/2} + \sum_{k=1}^{\infty} C_k(G^2t + a^2)^{-\lambda_k}$$

Here $C_0 > 0, C_k(k = 1, 2, ...)$ are the constants effectively calculated in terms of input data of problem (17)-(21), and $\lambda_k > 0$ are the eigenvalues of the Sturm-Liouville problem

$$\frac{d^2w}{dz^2} + \frac{z}{2}\frac{dw}{dz} + \lambda w = 0 \ \ \text{when} \ \ -G < z < 0$$

$$w(-G) = 0, \; w'(0) = 0$$

The other case when problem (17)-(20) admits an exact solution corresponds to the dependence

$$q = -\frac{G^2}{2\sqrt{(a^2 - G^2 t)}} \qquad (22)$$

The solution of the Stefan problem exists here only within the interval $0 \le t \le t_* = a^2/G^2$. At time $t = t_*$ the liquid phase occupy the region $-L < x < S(t)$ disappears. The function $S(t)$ in the solution of problem (17), (20), (22) may be represented as follows:

$$S = -L + d_0(a^2 - G^2 t)^{1/2} + d_1(a^2 - G^2 t)^{\mu_1} + \sum_{k=2}^{\infty} d_k(a^2 - G^2 t)^{\mu_k} \qquad (23)$$

Here $d_k(k = 0, 1, ...)$ are constants, $\mu_k > 0$ are the eigenvalues of the Sturm-Liouville problem,

$$\frac{d^2 w}{dz^2} - \frac{z}{2}\frac{dw}{dz} + \mu w = 0 \;\; when \;\; -G < z < 0$$

$$w(-G) = 0, \; w'(0) = 0$$

It is interesting to note that, according to (23), the main term of the asymptotics of the function $S(t) + L$ when $t \to t_{*-0}$ may be dependent on the parameter $G$. Namely, if $G < G_*$ where $G_* \approx 1.386$ is the unique positive kernel of the equation

$$\frac{G}{2} \int_0^G e^{y^2/4} dy = e^{G^2/4}$$

$\mu_1 > 1/2$ and $S + L = O(t_* - t)^{1/2}$ when $t \to t_{*-0}$. If $G > G_*, \mu_1 < 1/2$ and $S + L = O(t_* - t)^{\mu_1}$ when $t \to t_{*-0}$.

9. The Lagrangian coordinates may be successfully used not only in free-boundary problems, but also in problems with inner interfaces. One of them, Verigin problem, describes the process of filtration through a porous medium of two immiscible compressible liquids separated by a contact discontinuity surface. The liquids have different viscosities and equations of state $p_i = f_i(\rho_i), i = 1, 2$ which connect pressure r and density $\rho$. The filtration region is assumed to be finite, and either pressure or mass discharge is prescribed on its boundary. The density, velocity and pressure of each liquid satisfy the continuity equation and Darcy's law in a corresponding region. The pressure on the desired surface of contact discontinuity is continuous, and the normal velocity component in each

liquid coincides with the velocity of surface displacement in the formal direction to it.

In the case of one space variable, the Lagrangian coordinates allow transformation of the initial problem into an equivalent initial boundary-value one for a quasi-linear parabolic equation with fixed line of coefficient discontinuity. We succeeded to prove the existence and uniqueness of the classical solution of Verigin problem taking into account minimal assumptions on the basic data and to study qualitative properties of the solution [19, 20]. The existence of the solution in these works has been proved "in the large", for every $t > 0$, or up to the moment of a complete displacement of one liquid by another.

Obtaining of exact solutions of one-dimensional Verigin problem with a functional arbitrariness is of interest. Such solutions exist when the equation of state for each medium is especially chosen:

$$p_1 = C - \frac{A_1}{\rho_1^2}, \ p_2 = C \frac{A_2}{\rho_2^2}$$

Here $A_1, A_2$ and $C$ are positive constants, their availability is conditioned by the fact that for the above form of functions $f_i(\rho_i)$ not only differential equations but also the conditions of conjugation of their solutions on the contact discontinuity line unmovable on the plane $\xi, t$ are linearized after transformation of the problem to the Lagrangian coordinates.

10. The approach described above may be applied to the Cauchy problem with finite initial data for one-dimensional degenerate parabolic equations. The point is that the differential equation for the function determining the boundary of the solution support of the above equation may be interpreted to be a kinematic condition on the free boundary of the above-mentioned support [5, 8, 9]. As an example, let us consider the problem of determining the function from the conditions

$$u_t = \tau^{-(n-1)}[\tau^{n-1}(u^m)_z]_z \ when \ \tau, t > 0 \tag{24}$$

$$(u^{m-1})_z(0, t) = 0 \ when \ t > 0 \tag{25}$$

$$u(\tau, 0) = u_0(\tau) \geq 0 \ when \ \tau \geq 0 \tag{26}$$

Here $n$ is the natural number, $m > 1, u_0(\tau) > 0$ when $0 \leq \tau < a$ and $u_0 = 0$ when $\tau \geq a$. Besides, $u_0^{m-1} \in C^1[0, a], \ \tau^{-(1-n)}[\tau^{n-1}(u_0^{m-1})_z]_z \geq \gamma$ when $\tau \in [0, a]$ and $u_0 \geq (a - \tau)^\alpha$ when $\tau \to a - 0$ with some constants $\alpha, \gamma > 0$.

It is known [21] that when $n = 1$ the solution of problem (24)-(26) tends to some self-similar solution of equation (24), one-valued defined by the constant $m$ and by the value of the functional

$$M_1 = \int_0^\infty u_0(\tau) d\tau$$

It is also known that when $n = 1$ such a tendency is an asymptotical one [22]. Now let us come over to the mass Lagrangian coordinate in problem (24)-(26). We obtain the initial boundary-value problem for a quasi-linear parabolic equation which has been formulated in the half-band $\{0 < \xi < 1, t > 0\}$. It has been established [23] that for any natural $n$ the solution of problem (24)-(26) tends asymptotically to its self-similar limit determined by the constants $n, m$ and

$$M_n = \int_0^\infty \tau^{n-1} u_0(\tau) d\tau$$

In this solution, the equation of the function support boundary $u(\tau, t)$ is in the form

$$\tau = \eta(t) \equiv A t^{1/n(m-1+2/n)}[1 + O(t^{-1/n})]$$

when $t \to \infty$, where the constant $A > 0$ is expressed efficiently through $n$, $m$ and $M_n$.

11. The other example is associated with the Cauchy problem for the equation

$$u_t = (u^m)_{xx} + (u^\lambda)_x, \; m > 1, \; 0 < \lambda < 1 \tag{27}$$

arising in the filtration theory. The function $u_0$ entering the initial condition

$$u(x, 0) = u_0(x) \tag{28}$$

is such that $u_0 > 0$ for $x \in (0, a)$ and $u_0 = 0$ for $x \in \mathcal{R}(0, a)$. With some additional natural conditions imposed on this function, the theorem of existence and uniqueness of the generalized solution of problem (27), (28) has been proved [24].

Let us introduce the functions

$$\zeta(t) = inf\{x \in \mathcal{R} : u(x, t) > 0\}, \eta(t) = sup\{x \in \mathcal{R} : u(x, t) > 0\}$$

limiting the support of the generalized solution to the problem under consideration. As is shown in [24], under the conditions of the above-mentioned theorem, the equality $\zeta(t) = -\infty$ is valid for every $t > 0$. This property may be called the instant decompactification of the generalized solution support. For $\eta(t)$ the inequality $\eta(t) \le A_1 - A_2 t \; (A_1, A_2 = const. > 0)$ is valid in [24].

In [25] the author succeeded to obtain a more detailed information on a qualitative behaviour of the solution to problem (27) and (28). In [25] $\{u_0, (u_0^{m-\lambda})'\} \in C[0, a]$ and $u_0(x)$ supposed to be strictly monotonically decreasing functions when $x \to a - 0$, $x \to +0$. Besides, when $x$ is close to $a$, the function satisfies the differential inequality

$$(\frac{m}{m-1}(u_0^{m-1})' + u_0^{\lambda-1})' \ge -\beta$$

with some constant $\beta > 0$ (prime stands for differentiation with respect to $x$). Under these conditions the smoothness of the function $\eta(t)$ when $t \to 0$, its lower bound $\eta(t) \geq B_1 - B_2 t$ $(B_1, B_2 = const. >)$ as well as the equality

$$\lim_{x \to \eta(t)-0} (u^{m-\lambda})_x(x, t) = \frac{\lambda - m}{m}$$

have been found in [25]. Thus, the existence of the discontinuity points of the function $(u^{m-\lambda})_x$ is not associated with the smoothness of the function $u_0$; function $(u^{m-\lambda})_x$ and $u_x$ in the case of $m \geq 1 + \lambda$ undergo discontinuity at each point of the curve $x = \eta(t)$ when $t > 0$ This fact having no analogies in the theory of degenerate parabolic equations may be called the instant appearance of the singularity of such type solutions.

12. In conclusion let us consider the examples of nonlocally invariant solutions of degenerate parabolic equations which possess stronger singularities on the degeneration line as compared to the well-understood generalized solutions of similar equations. Our considerations refer to the family of equations of the form

$$w_t = [\frac{w^\alpha}{(\lambda w + \mu)^{\alpha+2}} w_x]_x \tag{29}$$

in which $\alpha$, $\lambda$ and $\mu$ are positive constants. Equation (29) degenerates when $w = 0$; we are interested in non-negative solutions of this equation. Here we shall restrict ourselves to the case $\alpha = 1$; as far as the parameters $\lambda$ and $\mu$ are concerned, they may be assumed equal to unity without loss of generality. Thus, let us consider the equation

$$w_t = [\frac{w}{(w+1)^3} w_x]_x \tag{30}$$

In (30) it is appropriate to substitute the required function $w = u - 1$ and then to apply L-transformation to the obtained equation for $u(x, t)$ . The corresponding equation for the function $\omega(\xi, t)$ is as follows: $\omega_t = [(1 - \omega)\omega_\xi]_\xi$. The last equation is reduced to the form

$$\chi_t = (\chi\chi_\xi)_\xi \tag{31}$$

due to the substitution of $\omega = 1 - \chi$. Eq.(31) possesses the four parameter Lie group in contrast to (30), where the admitted Lie group is a three parameter one [12]. Thus, the initial equation (30) possesses a hidden symmetry. This statement is also valid to the whole family of equations (29).

Equation (31) has a solution

$$\chi = \frac{\phi'(\xi)}{t+1} \tag{32}$$

where the function $\phi$ satisfies the equation

$$\phi'\phi'' + \phi = 0$$

Let us consider the even solutions of the last equation where $\phi' > 0$ for sufficiently small $\xi > 0$. They are determined by the quadrature

$$\xi = \int_0^\phi \frac{dz}{(A - 3z^2/2)^{1/3}}$$

where $A > 0$ is constant. Assume

$$\xi_* = \int_0^{\sqrt{2A/3}} \frac{dz}{(A - 3z^2/2)^{1/3}} \equiv 3\pi^{-1}\{\Gamma(2/3)\}^2(A/2)^{1/6}\}$$

Then $\phi' > 0$ for $0 \leq \xi < \xi_*$ and $\phi'(\xi_*) = 0$ The function $\phi(\xi)$ reaches its maximum when $\xi = \xi_*$ which is equal to $\sqrt{2A/3}$.

The solution (32) of (31) induces the connection between the Lagrangian and Eulerian variables

$$x = \xi - \frac{\phi(\xi)}{t+1} \tag{33}$$

The demand of mutual one-valuedness of (33) within $[0, \xi_*]$ for every $t \geq 0$ leads to the restriction $A \leq 1$. A non-negative solution $w(x,t)$ of (30) determined parametrically by equalities (33) and

$$w = \frac{\phi'(\xi)}{t + 1 - \phi'(\xi)} \tag{34}$$

where $|\xi| \leq \xi_*$, $t \geq 0$ corresponds to the invariant solution (32) of equation (31). Let us demand now that $0 < A < 1$ and assume $w(x,t) = 0$ for $|x| \geq \eta(t)$ where

$$\eta(t) = \xi_* - \frac{\phi(\eta_*)}{t+1}$$

is the functions defining the support boundary of the solution $w(x,t)$. Thereby, the function $ww_x$ has non-zero limits when $x \to -\eta(t)+0$ and $x \to \eta(t)-0$, $t \geq 0$. As a result, this solution is beyond the scope of a standard class of the generalized solutions to (30) and provides good grounds for calling it "the strange solution" of the above equation.

When $A \in (O, 1)$, the function $w(x,t)$ is a classical solution of equation (30) over the domain $|x| < \eta(t)$, $t > 0$. This function is continuous within a hemi-plane $t \geq 0$ of the plane $x, t$. As a result of a very strong singularity of the function $w$ at the points of the curve $x = \pm \eta(t)$ the conservation law

$$\frac{d}{dt} \int_{-\infty}^{\infty} w(x,t)dx = 0$$

taking place for ordinary generalized solutions of equation (30) becomes invalid, Therefore, it is difficult to attach physical sense to this solution if $w$ is identified as the density of some continuum (for example, to density of gas filtering through a porous medium;. One succeeded, however, to do it assuming $w$ to be the temperature not, density, of a medium undergoing phase transitions. Certainly, this approach is extraordinary, since with the transition temperature $w = 0$, the thermal conductivity coefficient of the medium vanishes.

As a consequence of (33), (34), the "differential equation" for $\eta(t)$ may take the form

$$\frac{d\eta}{dt} = - \lim_{x \to \eta(t) - 0} [(w + 1)^{-3} w w_x] \tag{35}$$

Relation (35) is equivalent to the condition on a free boundary $x = \eta(t)$ in the one-phase Stefan problem corresponding to heat-conduction equation (30) and to unit value of latent melting heat. It is interesting to note that when $t$ varies from 0 to $\infty$, the melting front $x = \eta(t)$ moves at a finite distance equal to $\phi(\eta_*) = \sqrt{2A/3}$. This qualitatively features the behaviour of the support boundary of the strange solution at high $t$ over its behaviour for the ordinary generalized solution of equation (30). Analogous strange solutions may be constructed for more general equation (29) when $\alpha, \lambda, \mu > 0$.

## REFERENCES

1. Pukhnachov V.V., Shlnareyov S.I. Investigation of quasilinear equations by the method of Lagrangian coordinates. Functional and numerical methods of mathematical physics. - Kiev: Naukova dumka, 1988, pp.181-185.

2 Meirmanov A.M., Pukhnachov V.V. Lagrangian coordinates in Stefan problem, Dinamika sploshnoi sredy. Novosibirsk, 47, 1980, pp. 90-111.

3. Shmareyov S.I. Self-similar solution of multi-dimensional Stefan problem, Dokl. Akad. Nauk SSSR, vol. 288, No. 1, 1986, p.95-99.

4. Pukhnachov V.V., Shmareyov S.I. Multi-dimensional self-similar Stefan problem in linearized statement, Reg. in VINITI No 4272-84, 1984, 25 p.

5. Gurtin M.E., McCamy R.C., Socolovsky E.A. A coordinate transformation that renders the free boundary stationary, Quart. J. Appl. Math., vol. 43, No. 3, 1984, pp.723-731.

6. Degtyarev L.M. , Krylov V.V. Numerical solution of the problem of wave field dynamics with singularities, Zhurnal vychislitel'noi matematiki i matematicheskoi fiziki, vol. 17, No: 6, pp.1523-1530.

7. Strampp W. Bäcklund transformation for diffusion equations, Physica D., vol. 6D, 1982, pp.113-118.

8. Pukhnachov V.V. Evolution equations and Lagrangian coordinates, Dinamika sploshnoi sredy. Novosibirsk, 70, 1985, pp.127-141.

9. Berryman J.C. Evolution of a stable profile for a class of nonlinear diffusion equations. 2.Slow diffusion on the line, J.Math. Phys., vol.21(6), 1980, pp.1326-1331.

10. Pukhnachov V.V. Equivalence transformations and hidden symmetry of evolution equations, Dokl. Akad. Nauk SSSR, vol.294, No: 3, 1987, pp.535-538.

11. Burgan J.R., Muder A., Reix M.R., Fialkov E. Homology and the nonlinear heat diffusion equation, SIAM J. Appl. Math., vol. 44, No. 1, 1984, pp.11-18.

12. Ovsiannikov L.V. Group properties of nonlinear heat condution equation, Dokl. Akad. Nauk SSSR, vol. 125, No.3, 1959, pp.492-495.

13. Akhatov I.Sh., Gazizov R.K., Ibragimov N.Kh. Group classification of nonlinear filtration equations, Dokl. Akad. Nauk SSSR, vol. 293, No. 5, 1987, pp. 1033-1035.

14. Akhmatov I.Sh., Gazizov R.K., Ibragimov N.Kh. Nonlocal symmetries. Euristic approach, Itogi nauki i tekhniki. Ser. Sovrelnennye problemy matematiki. Noveishie dostizheniia. vol. 34-M.: VINITI, 1989, pp.3-83.

15. Storm M.L. Heat conduction in simple metals, J. Appl. Phys., vol. 22, 1951, pp. 940-951.

16. Rosen G. Nonlinear heat conduction in solid H2, Rev. B2, vol. 19, No. 4, pp.2389-2399.

17. Dorodnitsyn V.A., Swirshchevsky S.R. On Lie-Bäcklund groups admitted by equation of heat-conduction with a source, Preprint, Keldysh Institute of Applied Mathematics, USSR Academy of Science, No. 101, 1983, 28p.

18. Focas A.S., Yortsos Y.C. On the exactly solvable equation $S_t = [(\beta S + \gamma)^{-2} S_x]_x + \alpha(\beta S + \gamma)^{-2} S_x$ occurring in two phase flow in porous medium, SIAM J. Appl. Math., vol 42, No. 2, 1982, PP-318-332.

19. Meirmanov A.M. On solvability of Verigin problem in exact statement, Dokl. Akad. Nauk SSSR, vol. 253, No. 3, 1980, pp. 588-591.

20. Meirmanov A.M. Problem of contact discontinuity surface motion at filtration of immiscible compressible liquids (Verigin problem), Sib. Mat. Zhurnal, vol. 23, No. 1, 1982, pp.85-102.

21. Friedman A., Kamin S. The asymptotic behaviour of gas in a n-dimensional porous medium, Trans. Amer. Mat. Soc., vol. 262, 1980, pp.551-563.

22. Vasquez J.L. Asymptotical behaviour and propagation properties of the one-dimensional flow of a gas in a porous medium, Trans. Amer. Math. Soc., vol. 277, 1983, pp.507-528.

23. Shmareyov S.I. Qualitative properties of generalized solutions of degenerate equations in filtration theory, Uspekhi matematicheskikhnauk, vol. 41 4(250), 1986, pp. 189-190.

24. Diaz J.I., Kersner R. Non-existence of one of the free boundaries in a degenerate equation in filtration theory, C. R. Acad. Sci. Paris, A 296, Ser. 1, 1983, pp.505-508.

25. Shmareyov S.I. Instantaneous appearance of the singularities in the solution of degenerate parabolic equation, Sibirskii matematicheskii zhurnal, vol. 31, No. 4, 1990, pp.166-179.

V. V. Pukhnachov
Lavrentyev Institute of Hydrodynamics
Siberian Division of the USSR Academy of Sciences,
Novosibirsk 630090, USSR

R RICCI

# Limiting behavior of some problems in reaction-diffusion

In this paper we consider some basic models for reaction-diffusion of fluids in solids. The foundamental variables will be fluid and solid reactant concentrations, denoted by $C_F$ and $C_S$ respectively. We will assume that a simple isothermal reaction takes place between the fluid and the solid reactant according to a reaction law expressed in terms of the *rate of reactions* $r_F$ and $r_S$ by

$$(1) \qquad \frac{dC_F}{dt} = r_F = a\, r_S = a\frac{dC_S}{dt},$$

where $a$ is the *stoichiometric coefficient* for the reaction. The rates of reaction are function of the local concentrations of the two reacting substances according to

$$(2) \qquad r_F = a\, r_S = -a\, kC_S^m\, C_F^n,$$

where $k$ is called *reaction rate constant* or *speed of reaction* and $m$ and $n$ are the *orders of the reaction*. All constants $a$, $k$, $m$ and $n$ are positive. In a more general non isothermal reaction the speed of reaction $k$ is no more a constant but depends, in general, on temperature.

If we consider a fluid diffusing into a porous solid matrix where a solid (immobile) reactant is fixed, then the raction law (1) holds at each point inside the solid and a balance law for the two concentrations must be written taking into account the flux of the fluid due to diffusion. Equations for the simplest model can be written assuming that the reaction does not affect the structure of the porous solid matrix, so that the effective diffusivity and porosity of the porous matrix are constant. Moreover in many relevant cases the products of the reaction can be neglected. According to these assumptions the equations for the model are

$$(3) \qquad \epsilon\, C_t - D\, \Delta C = -k\, C^n S^m, \quad \text{in } \Omega \times [0, T],$$

$$(4) \qquad S_t = -a\, k\, C^n S^m, \quad \text{in } \Omega \times [0, T],$$

where $\Omega$ is a bounded domain in $\mathbf{R}^d$, $d = 1, 2, 3$, $D$ is the effective diffusivity and $\epsilon$ the porosity of the solid, and we have simply denoted by $C$ the fluid concentration and by $S$ the solid reactant concentration.

This simple model can be usefully applied for describing ore reduction, retorting of oil shale, catalyst regeneration and deactivation [FB], [SS], [S]. We will call this model *Gas-Solid Model* in this paper.

Together with equations (3) and (4), initial conditions for both the solid reactant and the gas concentration in $\Omega$ and a boundary condition on $\partial\Omega \times [0, T]$ for the gas concentration must be given. We are mainly concerned with the following set of I.B.V. conditions

$$(5) \qquad\qquad C(x, 0) = 0, \quad \text{in } \Omega,$$

$$(6) \qquad\qquad S(x, 0) = S_0, \quad \text{in } \Omega,$$

$$(7) \qquad\qquad C(x, t) = C_0, \quad \text{on } \partial\Omega \times [0, T],$$

which represent the case in which the reactor $\Omega$ is initially void of gas, with uniformly distributed solid reactant and the gas concentration is kept constant (greater than 0) on the boundary of the reactor. This problem is called the *penetration problem*. If the order of the reaction for the gas, $n$, is less than 1 than a penetration front starts immediatly from $\partial\Omega$ and the gas concentration is positive between the exterior boundary and the penetration front, and vanishes inside the penetration front. Analogously if the reaction order for the solid, $m$, is less then 1 then an extinction front will appear after a finite time depending on $C_0$, $S_0$ and $m$, such that all the solid reactant is consumed between the extinction front and the exterior boundary, so that no reaction takes place in this region. Both fronts, if they exist, moves toward the interior of $\Omega$, invading the whole reactor in a finite time [S], [DH]. Let us recall some analytical results for equations (3)-(7). The results are better stated introducing the variable

$$X(x, t) = S_0 - S(x, t)$$

so that the system reads now

$$(8) \qquad\qquad \epsilon\, C_t - D\, \Delta C = -k\, F(C, X), \quad \text{in } \Omega \times [0, T],$$

$$(9) \qquad\qquad X_t = a\, k\, F(C, X), \quad \text{in } \Omega \times [0, T],$$

$$(10) \qquad\qquad C(x, 0) = 0, \quad \text{in } \Omega,$$

$$(11) \qquad\qquad X(x, 0) = 0, \quad \text{in } \Omega,$$

$$(12) \qquad\qquad C(x, t) = C_0, \quad \text{on } \partial\Omega \times [0, T],$$

where $F(C, X) = C^n(S_0 - X)^m$.

The reaction vector $(-kF, akF)$ is now monotone in off-diagonal terms, i.e. $-kF$ is monotone increasing with respect to $X$, and $kaF$ is monotone increasing with respect to $C$. This allows to define super and subsolutions for this system substituting the $=$ sign in (8) and (9) by $\geq$ and $\leq$ respectively. Then a comparison theorem holds saying that if $(\overline{C}, \overline{X})$ and $(\underline{C}, \underline{X})$ are a supersolution and a subsolution such that $\overline{X} \geq \underline{X}$ in $\Omega \times \{t = 0\}$ and $\overline{C} \geq \underline{C}$ on the parabolic boundary $\Omega \times \{t = 0\} \cup \partial\Omega \times (0, T)$ then $\overline{X} \geq \underline{X}$ and $\overline{C} \geq \underline{C}$ in $\Omega \times (0, T)$ (the inequalities are strict if at lest one of the above inequalities is and both the orders of reaction are greater or equal to 1, on the contrary, if at least one of $m$ or $n$ is less than one the solutions may have a coincidence set even if they are strictly ordered on the parabolic boundary).

The uniqueness of the solution follows immediatly from the comparison principle. Moreover a topological existence theorem can be proved, namely if there exist a supersolution $(\overline{C}, \overline{X})$ and subsolution $(\underline{C}, \underline{X})$ such that $\overline{C} \geq \underline{C}$ and $\overline{X} \geq \underline{X}$, then a solution $(C, X)$ exists, such that $\overline{C} \geq C \geq \underline{C}$ and $\overline{X} \geq X \geq \underline{X}$. For the penetration problem obvious super and subsolutions are $(1, 1)$ and $(0, 0)$, which ensure the existence of a unique solution of the problem [P], [S], [DS].


Another model is often used in application. Now we assume that the solid is essentialy impervious to the gas penetration and then the reaction can only take place on a narrow region called *reaction surface* . As a consequnce of the reaction the solid changes its structure giving rise to a porous inert product generally called *ash*. In the ash the gas can diffuse to reach a new location of the reaction surface pushing it inside the impervious solid. This model is known under the name of *Unreacted Core Model*. The mathematical model describing this process is a free boundary problem for the diffusion equation which rules the fluid distribution inside the ash. The free boundary is obviously the reaction surface. On this surface the reaction is assumed to obey to a localized version of the reaction laws (1) and (2). Let us write a mathematical model assuming planar symmetry. Then we assume that the ash occupies the slab $0 < x < s(t)$, while the unreacted solid fill the half space $x > s(t)$, $x = s(t)$ giving the location of the reaction front at time $t$. In the ash layer the gas diffuses according to the Fick's law, so that

$$(13) \qquad \epsilon\, c_t - Dc_{xx} = 0, \quad 0 < x < s(t),\ t > 0,$$

where $c$ indicates the gas concetration, $\epsilon$ the ash porosity and $D$ the effective diffusivity of the gas in the ash.

Since the amount of consumed solid reactant per unit of time is given by $c_S \frac{ds}{dt}$, $c_S$ indicating the solid reactant concetration, the reaction law at the reaction surface becomes

$$(14) \qquad \frac{ds}{dt}(t) = kc_S^{m-1}c^n(s(t), t), \quad t > 0,$$

We assume in the following that $c_S$ is constant, i.e. the solid is initially homogeneous.

A second condition is needed on $x = s(t)$, since the function $s(t)$ is not known a priori. This condition is given by the local balance of mass at the front. Since from (14) $c(s(t), t)$ is aspected to be greater than 0, then $c(x, t)$ is discontinuous across the free boundary, and a transport term $\epsilon c(s(t), t) \frac{ds}{dt}(t)$ will appear in the balance law from the Rankine-Hugoniot jump condition. Moreover part of the gas is consumed at the front to push forward the front itself. The amount of this consumed gas, given by the the reaction law, is $akc_S^m c^n(s(t), t)$. The gas is supplied at the free boundary from diffusion inside the ash layer, so finally the mass balance equation becomes

$$(15) \qquad -D\frac{\partial}{\partial x}c(s(t), t) = akc_S^m c^n(s(t), t) + \epsilon c(s(t), t)\frac{ds}{dt}(t), \quad t > 0.$$

The penetration problem now is given by equations (13)-(15) with initial and boundary conditions

$$(16) \qquad s(0) = 0,$$

$$(17) \qquad c(0, t) = c_0, \quad t > 0.$$

which mean that initially no ash is present, and the concetration is kept constant at the exterior boundary.

The same system of equations arises from a model describing the penetration of solvents into glassy polymers [AS],[A], [FR]. For many couples solvent-polymer the sorption of the solvent by the polymer causes a change in the structure of the polymer itself, basically due to desagration of long macromolecules chains, that transforms the polymer into a gel. This transformation only takes place for value of the solvent concentration exceding a particulat threshold value which is typical of the couple solvent-polymer and may also depends on temperature and pressure. The transition zone from the glassy state to the gel is very sharp, and is assimilated to a surface in this model. Inside the gel the solvent diffuses from the exterior toward the transition front, while the glassy zone is assumed to be unpenetrated by the solvent. At the front mass conservation holds (with a Rankine-Hugoniot jump condition since the solvent is dicontinuous across it) and a kinematical law relating the speed of penetration of the front to the local value of the solvent concetration is assumed. This penetration law is inspired by the reaction law in the case of the Unreacted Core Model, giving rise to a free boundary condition similar to (14). The complete set of equations in the penetration problem is now, in normalized variables,

$$(18) \qquad c_t - c_{xx} = 0, \quad 0 < x < s(t), \ t > 0,$$

$$(19) \qquad s(0) = 0,$$

$$(20) \qquad c(0, t) = 1, \quad t > 0,$$

$$(21) \qquad c_x(s(t), t) = -\frac{ds}{dt}(t)c(s(t), t), \quad t > 0,$$

$$(22) \qquad \frac{ds}{dt}(t) = \beta\left[c(s(t), t) - q\right]^n, \quad t > 0,$$

where $c$ is the total normalized concetration and $q < 1$ the normalized threshold value. The constant $\beta$ is proportional to the inverse of a relaxation time which is typical of the couple solvent-polymer [A].

If we rewrite this system in terms of the excess of concentration $C = c - q$, then we get the same mathematical problem we have for the Unreacted Core Model. Other types of boundary contitions at $x = 0$ can also be of interest.

Analitical resuts for these problems can be found in [FMP], [AR], [CR1], [CRT]. In particular for the penetration problem with boundary condition (17), existence and uniqueness of the solution has been proved in [FMP], assuming a general penetration law of the form

$$(22') \qquad \frac{ds}{dt}(t) = f(c(s(t), t) - q), \quad t > 0,$$

where $f \in C[0, 1] \cap C^1(0, 1]$, $f(0) = 0$. Under this assumption the free boundary satifies $s \in C^1[0, +\infty) \cap C^2(0, +\infty)$, and $s \in C^\infty(0, +\infty)$ if $f$ is $C^\infty$. Moreover $\frac{ds}{dt} > 0$ and $\frac{d^2 s}{dt^2} < 0$. For $t \to \infty$ we have the following asymptotic estimates of the free boundary:

$$(23) \qquad s(t) \geq \sqrt{\frac{2t}{q + \frac{1}{3}}}\left(1 + e^2(t)\right), \quad e \to 0 \text{ as } t \to \infty,$$

which holds for any $q \geq 0$. Moreover if $q > 0$ then an estimate from above is also available:

$$(24) \qquad s(t) < \sqrt{\frac{2t}{q}}.$$

## THE PSEUDO STEADY STATE APPROXIMATION

A common approximation for the Gas-Solid Model as well as for the Unreacted Core Model is the so called Pseudo Steady State Approximation, PSSA, which consists in replacing the diffusion operator by the Laplace operator. This is justified from a physical point of view by the fact that the porosity $\epsilon$ is often very small. A mathematical justification of this procedure must give an estimate of the difference of the two solutions, $\epsilon > 0$ and $\epsilon = 0$, in some appropriate function space. For the Gas-Solid Model this problem was solved in [SBG], see also [S], [DS]. Let us briefly recall this result. The PSSA here is given by the system

$$(25) \qquad -D\,\Delta C = -k\,C^n S^m, \quad \text{in } \Omega \times [0, T],$$

$$(26) \qquad S_t = -a \, k \, C^n S^m, \quad \text{in } \Omega \times [0,T],$$

$$(27) \qquad S(x,0) = S_0, \quad \text{in } \Omega,$$

$$(28) \qquad C(x,t) = C_0, \quad \text{on } \partial\Omega \times [0,T],$$

where the initial condition on the gas concetration $C$ has been dropped since (25) is an elliptic equation. This suggests that any reasonable norm to estimate the difference between the solution of (25)-(28), let us denote it simply by $(C,S)$, and the solution of (3)-(7), denoted by $(C_\epsilon, S_\epsilon)$, must take into account that the difference between $C(x,0)$ and $C_\epsilon(x,0)$ is out of control since $C(x,0)$ is determined by the boundary condition and the initial condition $S(x,0)$, while $C_\epsilon(x,0)$ is arbitrary (if we consider the general problem and not simply the penetration problem). In fact a good estimate, assuming that $C_\epsilon(x,0) \leq C(x,0)$ (this is the case for the penetration problem) and $S_\epsilon(x,t) = S(x,t)$, is given by

$$(29) \qquad 0 \leq \int_0^t (C(x,\tau) - C_\epsilon(x,\tau)) \; d\tau \leq Const.(\Omega) \, \epsilon,$$

$$(30) \qquad 0 \leq \int_\Omega (S(x,t) - S_\epsilon(x,t)) \; dx \leq Const.(\Omega) \, \epsilon,$$

where the constants in the estimates depends on the domain $\Omega$, [**DS**].

The same limit $\epsilon \to 0$ for the Unreacted Core Model gives rise to the system

$$(31) \qquad c_{xx} = 0, \quad 0 < x < s(t), \; t > 0,$$

$$(32) \qquad \frac{ds}{dt}(t) = k c_S^{m-1} c^n(s(t),t), \quad t > 0,$$

$$(33) \qquad -D \frac{\partial}{\partial x} c(s(t),t) = a k c_S^m c^n(s(t),t), \quad t > 0.$$

$$(34) \qquad s(0) = 0,$$

$$(35) \qquad c(0,t) = c_0, \quad t > 0.$$

In this case the problem can be easily reduced to a single nonlinear ordinary differential equation for the free boundary $s(t)$, namely

$$(36) \qquad \left(\frac{ds}{dt}\right)^{\frac{1}{n}} + \frac{A}{D}K^{\frac{1}{n}}s\frac{ds}{dt} = \frac{K}{n}, \quad t > 0,$$

$$(37) \qquad s(0) = 0,$$

where $A = akc_S^m$ and $K = kc_S^{m-1}$.

We indicate again by $(c, s)$ the solution of (31)-(35) and by $(c_\epsilon, s_\epsilon)$ the solution of (13)-(17). The difference between the two solutions can now be mesured estimating the $L^\infty$ norm of the diffence $s - s_\epsilon$, [CR2].

This can be done writing an equation for $s_\epsilon$ similar to (36). To do that we start from the Green's identity for the heat equation, which gives an alternative free boundary condition

$$(38) \qquad \frac{1}{2}As_\epsilon^2(t) + D\int_0^t c_\epsilon(s_\epsilon(\tau), \tau)d\tau = Dt - \int_0^{s_\epsilon(t)} \epsilon x c_\epsilon(x, t)dx.$$

Differentiating (38) and isolating the terms involving $c_\epsilon$, we get

$$(39) \qquad \left(\frac{ds}{dt}\right)^{\frac{1}{n}} + \frac{A}{D}K^{\frac{1}{n}}s\frac{ds}{dt} = \frac{K}{n} + \epsilon\Gamma_\epsilon(t),$$

where

$$(40) \qquad \Gamma_\epsilon(t) = s_\epsilon(t)\frac{ds}{dt}(t)c_\epsilon(s_\epsilon(t), t) + \int_0^{s_\epsilon(t)} c_{\epsilon t}(x, t)dx.$$

Uniform estimates on $c_\epsilon$, $s_\epsilon$ and $\frac{ds}{dt}$, [CR2], imply that the function $\Gamma_\epsilon$ is positive and is dominated by a function only depending on $t$, uniformly with respect to $\epsilon$. Then Gronwall's lemma implies that

$$(41) \qquad 0 < s(t) - s_\epsilon(t) < Const.(T) \, \epsilon, \quad 0 < t < T.$$

An estimate of the same type, but with a better constant in the inequality (41), has been recently obtained in [H], using a completely different approach. In this paper the dependence of the constant on time is of the order $\sqrt{T}$, which agrees with the fact that the domains of the two equations (13) and (31) increase with time like $\sqrt{T}$.

## The Fast Reaction Approximation

The second type of approximation we want to consider is the case of very large *speed of reaction k*. We still consider reaction of the type described by (1) and (2), but now we let $k \to \infty$. From the point of view of the spatially omogeneous reaction decribed by (1), only the two trivial equilibrium states are possible, namely $C_F = 0$ or $C_S = 0$. So the expected asymptotic problem for the spatially distibuted reaction will be a free boundary problem, with a jump in the non diffusing species. A similar asymptotic limit can be done for the Unreacted Core Model. This is itself a free boundary problem. However the limiting problem will have a different free boundary condition. In fact, in order to keep bounded the front speed in (14), one must assume that the gas concentration vanishes at the free boundary. Then condition (15) becomes the standard one-phase Stefan condition. The same limit in the polymer penetration model can be interpreted as taking the limit of the relaxation time going to 0.

Similar limits have been considered for related models. Let us mention in particular the Gas-Gas Reaction model [CH], [E], the Stefan problem with relaxation [V], and the Phase Field model [C].

The Gas-Gas Reaction model is very similar to the Gas-Solid model we have considered. Instead of gas reacting with a solid reactant fixed to a porous matrix, here two gaseous species are diffusing inside a porous matrix, and undergo a reaction according to (1), (2). Then the basic equations are, in normalizied variables $u$ and $v$ indicating the two gas concetrations and assuming that both order of reaction are equal to 1,

$$(42) \qquad u_t - \Delta u = -\lambda k u v,$$

$$(43) \qquad v_t - \Delta v = -k u v,$$

both holding in a bounded domain $\Omega \subset \mathbf{R}^d$ and for $t \in (0, T)$.

In the limit $k \to \infty$ the concetration will satisfy $uv = 0$ and equations (42) and (43) are replaced by

$$(44) \qquad u_t - \Delta u = 0, \quad \text{in } \Omega_u,$$

$$(45) \qquad v_t - \Delta v = 0, \quad \text{in } \Omega_v,$$

where $\Omega_u = \{(x, t) : u(x, t) > 0\}$, $\Omega_v = \{(x, t) : v(x, t) > 0\}$, and free boundary conditions on the interface $\Gamma = \partial\Omega_u \cap \partial\Omega_v$

$$(46) \qquad u(x, t) = v(x, t) = 0, \quad (x, t) \in \Gamma \times (0, T),$$

$$(47) \qquad \lambda\frac{\partial u}{\partial \underline{n}} + \frac{\partial v}{\partial \underline{n}} = 0, \quad (x, t) \in \Gamma \times (0, T),$$

Problem (44)-(47) was studied in [CR] and the convergence of the solution of (42)-(43) to the solution of the free boundary problem as $k \to \infty$ was proved in [E] by means of a compactness argument.

In the Stefan problem with phase relaxation considered in [V] a new variable $\chi$, representing the local phase content, is introduced and an evolution equation for this new variable is coupled with the energy balance equation. The resulting system is then

$$(48) \qquad u_t + L\chi_t = K\Delta u,$$

$$(49) \qquad \beta(u) \in \epsilon \chi_t + H^{-1}(\chi),$$

where $u$ denotes the temperature, $\beta$ is a monotone function, $L$ is the latent heat and $H$ is the Heaviside graph. The small parameter $\epsilon$ is a relaxation time, and (49) is a relaxed version of the relation between temperature and phase content in the well known enthalpy formulation of the Stefan problem [EO]. The mathematical problem arising from (48)-(49) has been studied in [V], where also the convergence in the limit $\epsilon \to 0$, to the solution of the Stefan problem has been proved.

Similarly in the Phase Field model a phase function $\phi$ is introduced. In this model the evolution equation for the phase is a diffusion equation coupled with the energy conservation

$$(50) \qquad u_t + \frac{l}{2}\phi_t = K\Delta u,$$

$$(51) \qquad \alpha\xi^2\phi_t = \xi^2\Delta\phi + a^{-1}g(\phi) + 2u.$$

Using formal asymptotic expansion for different possible limits on the five parameters involved in the model one can recover the Stefan problem and some modified Stefan and Hele Shaw problems, [C]. However rigorous convergence results are not available for these limits.

Let us start considering the limit for infinite speed reaction in the Unreacted Core model or, equivalently from the mathematical point of view, for the time relaxation going to 0 in the polymer penetration problem. We rewrite equations (18)-(22) in terms of the excess of concentration, restricting for semplicity to the linear penetration law, $n = 1$. We denote by $C_\beta$ and $s_\beta$ the solution of the problem for $\beta < \infty$ and simply by $C$ and $s$ the solution of the limiting problem for $\beta = \infty$. Then $(C_\beta, s_\beta)$ solves

$$(52) \qquad C_{\beta t} - C_{\beta xx} = 0, \quad 0 < x < s(t), \ t > 0,$$

$$(53) \qquad s_\beta(0) = 0,$$

$$(54) \qquad C_\beta(0, t) = 1, \quad t > 0,$$

86

$$(55) \qquad C_{\beta x}(s_\beta(t), t) = -\frac{ds_\beta}{dt}(t)\left[C_\beta(s_\beta(t), t) + q\right], \quad t > 0,$$

$$(56) \qquad \frac{ds_\beta}{dt}(t) = \beta C_\beta(s_\beta(t), t), \quad t > 0,$$

whether $(C, s)$ is the solution of (52)-(54) and

$$(57) \qquad C(s(t), t) = 0, \quad t > 0,$$

$$(58) \qquad C_x(s(t), t) = -q\frac{ds}{dt}(t), \quad t > 0.$$

First notice that for $q = 0$ the limit problem does not make sense. In fact, for any solution of (52)-(54) and (57) with a given boundary $x = s(t) > 0$, points on the boundary $x = s(t)$ are minimum points and then $C_x(s(t), t) < 0$ from the boundary point principle [**F1**], thus contradicting (58). The case $q = 0$ can not be a priori neclected in the polymer penetration problem (for the Unreacted Core $q$ is always positive). In this case the limiting behavior is simple diffusion with no free boundary.

In order to prove the convergence of $s_\beta$ to $s$ we introduce the cumulative concentration $u(x, t)$ defined by

$$(59) \qquad u(x, t) = \int_x^{s(t)} [C(\xi, t) + q]\, d\xi,$$

and similarly $u_\beta(x, t)$ starting from $C_\beta$ and $s_\beta$.

The transformation (59), introduced in [**FMP**], reduces problem (52)-(56) to

$$(60) \qquad u_{\beta t} - u_{\beta xx} = 0, \quad 0 < x < s(t), \ t > 0,$$

$$(61) \qquad s_\beta(0) = 0,$$

$$(62) \qquad u_{\beta x}(0, t) = -1 - q, \quad t > 0,$$

$$(63) \qquad u_\beta(s_\beta(t), t) = 0, \quad t > 0,$$

$$(64) \qquad -u_{\beta x}(s_\beta(t), t) = \frac{1}{\beta}\frac{ds_\beta}{dt}(t) + q, \quad t > 0,$$

and similarly $(u(x,t), s(t))$ solves (60)-(63) and

(64') $$-u_x(s(t), t) = q, \quad t > 0.$$

Taking the limit $\beta \to \infty$ in (60)-(64) when $q = 0$ is the same as taking the limit of the latent heat tending to 0 in the Stefan problem. The result is diffusion in the half space, with no free boundary, [SH].

We recall some properties of the solutions of these problems, [FPR]. The dependence of the free boundaries on the parameter $\beta$ is monotone, i.e. if $(u_i, s_i)$ are solutions corresponding to values $\beta_i$, $i = 1, 2$ then $\beta_1 < \beta_2$ implies $s_1(t) < s_2(t)$. Moreover for any $\beta$ we have

(65) $$s_\beta(t) < s(t), \quad t > 0,$$

(66) $$u_\beta(x,t) < u(x,t), \quad 0 < x < s_\beta(t), \quad t > 0,$$

and finally

(67) $$u(x,t) > q(s(t) - x), \quad 0 < x < s_\beta(t), \quad t > 0,$$

From these estimates we get the convergence of $s_\beta$ to $s$, [FPR].

THEOREM. *The following estimate holds*

(68) $$0 < s(t) - s_\beta(t) < \frac{Const.(T)}{q^{\frac{1}{2}}} \left(\frac{1}{\beta}\right)^{\frac{1}{2}}, \quad 0 < t < T,$$

*where the constant in (68) is $O\left(T^{\frac{1}{2}}\right)$ as $T \to 0$ and $O\left(T^{\frac{1}{4}}\right)$ as $T \to \infty$.*

PROOF. Integrating the heat equation one obtains

$$\frac{1}{\beta} s_\beta(t) = t - \int_0^{s_\beta(t)} u_\beta(x,t)dx,$$

and similarly for $\beta = \infty$

$$0 = t - \int_0^{s(t)} u(x,t)dx.$$

Hence, using (65) and (66) we have

$$\frac{1}{\beta} s_\beta(t) > t - \int_{s_\beta(t)}^{s(t)} u(x,t)dx,$$

and then, by (67)

$$\frac{2}{\beta} s_\beta(t) > q \left( s(t) - s_\beta(t) \right)^2,$$

from which (68) follows. The asympototic evaluation of the constant in (68) are a consequence of the estimates for $s_\beta$ in [**FMP**].

Thi result can be easily extended to any reaction term $F(c) = \beta \Psi(c)$, where $\Psi$ is any strictly monotone smooth function, $\Psi(0) = 0$, [**FPR**].

In order to discuss the limiting behavior for the Gas-Solid Model we first recall the variational setting for the one-phase Stefan problem in more than one spatial dimension.

Let $\Omega$ be a bounded domain in $\mathbf{R}^d$ with regular boundary $\partial\Omega$. The one-phase Stefan problem consists in finding a function $\tau(x)$ defined in $\Omega$ and a positive function $c(x,t)$ defined for $t > \tau(x)$, vanishing elsewhere in $\Omega \times (0,T)$, such that

$$(69) \qquad c_t - \Delta c = 0, \quad x \in \Omega, \ \tau(x) < t < T,$$

$$(70) \qquad c(x,0) = 0, \quad x \in \Omega,$$

$$(71) \qquad c(x,t) = 1, \quad x \in \partial\Omega, \ 0 < t < T,$$

$$(72) \qquad c(x,\tau(x)) = 0, \quad x \in \Omega,$$

$$(73) \qquad \nabla\tau(x) \cdot \nabla c(x,\tau(x)) = -\lambda, \quad x \in \Omega,$$

where we have chosen very special initial and boundary conditions, which will correspond to the penetration problem for the Gas-Solid Model in the limit $k \to \infty$. From the point of view of the standard interpretation of the Stefan problem in terms of temperature, these initial and boundary conditions correspond to a block of ice, initially at the melting temperature, which is melted imposing at the outer boundary of the water-ice system a fixed positive temperature.

It is well known that problem (69)-(73) can be reformulated in term of a variational inequality, see for instance [**F2**], introducing the function

$$(74) \qquad U(x,t) = \begin{cases} \int_{\tau(x)}^t c(x,y)dy, & x \in \Omega, \ t > \tau(x), \\ 0, & x \in \Omega, \ t < \tau(x). \end{cases}$$

Then the function $U(x,t)$ satisfies

$$(75) \qquad U(x,0) = 0, \quad x \in \Omega,$$

$$(76) \qquad\qquad U(x,t) = t, \quad x \in \partial\Omega,\ 0 < t < T,$$

and

$$(77) \qquad\qquad U_t - \Delta U + \lambda = 0, \quad \text{where } U > 0,$$

$$(78) \qquad\qquad U \geq 0, \quad \text{in } \Omega \times (0,T).$$

These two conditions can be written concisely saying that $U$ is the (nonnegative) solution of

$$(79) \qquad\qquad U_t - \Delta U + \lambda H(U) = 0,$$

where $H$ is the Heaviside function

$$H(u) = \begin{cases} 1, & u > 0 \\ 0, & u \leq 0. \end{cases}$$

The standard approach to solve equation (79) is to introduce a sequence of smooth functions $H_\epsilon$ which approximate in an approprite way the function $H$. This is known as the penalization method. The penalizing sequence is such that

$$\begin{aligned}
&H_\epsilon'(u) \geq 0, \\
&H_\epsilon(u) \to -\infty, \quad \text{if } u < 0,\ \epsilon \to 0, \\
&H_\epsilon(u) \uparrow 1, \quad \text{if } u > 0,\ \epsilon \to 0, \\
&H_\epsilon(0) = 0.
\end{aligned}$$

Then we replace (79) by

$$(80) \qquad\qquad U_t - \Delta U + \lambda H_\epsilon(U) = 0,$$

Solving these penalized equations we get a sequence of $U_\epsilon$ converging to the solution of the variational Stefan problem in the limit $\epsilon \to 0$.

Let us came back to the Gas-Solid Model system choosing as initial and boundary conditions those for the penetration problem. Then, setting all constants which are mathematically irrelevant in this limit equal to 1, and restricting to the case $n = 1$, i.e. the reaction term is linear in the gas concentration $C_k$, we have to solve

$$(81) \qquad\qquad C_{kt} - \Delta C_k = -\lambda k C_k S_k^m, \quad x \in \Omega,\ 0 < t < T,$$

$$(82) \qquad\qquad S_{kt} = -k C_k S_k^m, \quad x \in \Omega,\ 0 < t < T,$$

(83)
$$C_k(x,0) = 0, \quad x \in \Omega,$$

(84)
$$S_k(x,0) = 1, \quad x \in \Omega,$$

(85)
$$C_k(x,t) = 1, \quad x \in \partial\Omega, \ 0 < t < T,$$

Now we introduce the new variable

(86)
$$U_k(x,t) = \int_0^t c(x,y)dy, \quad x \in \Omega, \ 0 < t < T.$$

We can now rewrite (81) substituting the reaction term by the time derivative of the solid reactant concentration, $C_{kt} - \Delta C_k = \lambda S_{kt}$, and integrate with respect to time. Taking into account the initial condition for $C_k$ and $S_k$, we find that $U_k$ and $S_k$ must solve

(87)
$$U_{kt} + \lambda(1 - S_k) - \Delta U_k = 0, \quad x \in \Omega, \ 0 < t < T.$$

On the other hand equation (82) can be integrated in terms of $U_k$, getting $S_k(x,t) = A_k(U_k(x,t))$, where

(88)
$$A_k(u) = \begin{cases} [1 - (1-m)ku]_+^{\frac{1}{1-m}}, & \text{if } m < 1, \\ \exp(-ku), & \text{if } m = 1, \\ [1 - (1-m)ku]^{\frac{1}{1-m}}, & \text{if } m > 1. \end{cases} .$$

Finally, we define

(89)
$$H_k(u) = 1 - A_k(u),$$

and we substite into (87) and we get an equation for the variable $U_k$ alone.

(90)
$$U_{kt} - \Delta U_k + \lambda H_k(U_k) = 0,$$

It easy to realize that equation (90) is a penalization of equation (79), and then the sequence of the solutions of (90), for different values of $k$, converges to the solution of (79) as $k \to \infty$.

Since we have an explicit form of the penalization term we can also evaluate the rate of convergence of $U_k$ to $U$ in the $L^\infty(\Omega)$ norm. First notice that the the comparison principle implies that the sequence $U_k$ is monotone decreasing, so that ([FPR])

(91)
$$0 \leq U(x,t) < U_k(x,t).$$

91

LEMMA. ([**FPR**]) *Let $U$ be the solution of (79) and $U_\epsilon$ the solution of (80) (with the same initial and boundary conditions). Then for any function $\delta(\epsilon)$ such that $\delta(\epsilon) \downarrow 0$ as $\epsilon \to 0$ we have*

(92) $$0 < U_\epsilon - U < \delta(\epsilon) + T\lambda(1 - H_\epsilon(\delta(\epsilon))), \quad \text{in } \Omega \times (0,T).$$

PROOF. Define $V_\epsilon(x,t) = U_\epsilon(x,t) - U(x,t)$. Then $V_\epsilon > 0$ and solves

$$V_{\epsilon t} - \Delta V_\epsilon = f_\epsilon(x,t) = \lambda(H(U(x,t)) - H_\epsilon(U_\epsilon(x,t)))$$

with $V_\epsilon = 0$ in $\Omega \times (0,T)$ and on $\partial\Omega \times (0,T)$.

The source term $f_\epsilon$ allways nonnegative and it is positive when $u > 0$.

Now take any function $\delta(\epsilon)$, decreasing to 0 with $\epsilon$.

Let $\Omega_\epsilon = \{(x,t) : V_\epsilon(x,t) > \delta(\epsilon)\}$, then either $\Omega_\epsilon$ is void, i.e. $V_\epsilon < \delta(\epsilon)$ in $\Omega \times (0,T)$ and (92) is trivially satisfied, or

$$f_\epsilon(x,t) < \lambda(1 - H_\epsilon(\delta(\epsilon))), \quad \text{in } \Omega_\epsilon,$$

and so $V_\epsilon$ is dominated in $\Omega \times (0,T)$ by the solution $z$ of

$$z_t - \Delta z = \lambda(1 - H_\epsilon(\delta(\epsilon))),$$

$$z(x,0) = \delta(\epsilon), \quad x \in \Omega,$$

$$z(x,t) = \delta(\epsilon), \quad (x,t) \in \partial\Omega \times (0,T).$$

Since $z(x,t)$ is trivially dominated by $\delta(\epsilon) + T\lambda(1 - H_\epsilon(\delta(\epsilon)))$, inequality (92) follows immediately.

We can now choose the function $\delta(\epsilon)$ in (92) to get the best possible estimate estimate of the difference $U_\epsilon - U$.

If the order of the reaction for the solid $m = 1$, then $H_k(u) = 1 - \exp(-ku)$, and taking $k = \frac{1}{\epsilon}$ and $\delta(\epsilon) = \epsilon \log \frac{1}{\epsilon}$, we get an order $(\frac{1}{k}\log(k))$ convergence of our approximate solution $U_k$ to $U$ in the $L^\infty$ norm.

If the order $m < 1$, then the penalized functions $H_k(u)$ are identically equal to 1 for $u > \frac{1}{k(1-m)}$. Then, taking $\delta(\epsilon) = \frac{\epsilon}{1-m}$, we get an order $\frac{1}{k}$ convergence. Similarly in the case $m > 1$, we get an order $\left(\frac{1}{k}\right)^{\frac{1}{m}}$ convegence.

REFERENCES

[AR] D. Andreucci, R. Ricci. A free boundary problem arising from sorption of solvents in glassy polymers. *Quarterly Appl.Math.*, **44** (1987), 649-657.

[A] G. Astarita. A class of free bondary problems arising in the analysis of transport phenomena in polymers, in *Free boundary problems: Theory and applications*, A. Fasano, M. Primicerio ed.s, Reserch notes in Math. **79** , Pitman, 1983.

[AS] G. Astarita, G.C. Sarti. A class of mathematical models for sorption of swelling solvents in galssy polymers. *Polym.Eng.Sci.*, **18** (1978), 388-395.

[BS] C. Bandle, I. Stakgold. The formation of the dead core in parabolic reaction-diffusion equations. *Trans. Amer. Math. Soc.*, **286** (1984), 275-293.

[C] G. Caginalp. Stefan and Hele-Shaw type models as asymptotic limits of the phase-field equations. *Phys.Rew.A*, **39** (1989), 5887-5896.

[CH] J.R. Cannon, C.D. Hill. On the movement of a chemical reaction interface. *Indiana Math. J.*, **20** (1970), 429-454.

[CR1] E. Comparini, R. Ricci. On the swelling of a glassy polymer in contact with a well-stirred solvent. *Math.Meth. in Appl.Sciences*, **7** (1985), 238-250.

[CR2] E. Comparini, R. Ricci. Convergence to the pseudo-staedy-state approximation for the unreacted core model. *Applicable Analysis*, **26** (1988), 305-325.

[CRT] E. Comparini, R. Ricci, C. Turner. Penetration of a solvent into a nonhomegeneous polymer. *Meccanica*, **23** (1988), 75-80.

[DH] J.I. Diaz, J. Hernandez. On the existence of a free boundary for a class of reaction diffusion systems. *SIAM J.Math.Anal.*, **15** (1984), 670-685.

[DS] J.I. Diaz, I. Stakgold. Gas-solid reaction, to appear.

[EO] C. Elliott, J.R. Ockendon. *Weak an variational methods for moving boundary problem*, Reseach notes in Math. **59**, Pitman, 1982.

[E] L.C. Evans. A convergence theorem for a chemical diffusion-reaction system. *Houston J.Math.*, **6** (1980), 259-267.

[FMP] A. Fasano, G. Mayer, M. Primicerio. On a problem in polymer industry: Theoretical and numerical investigation . *SIAM J.Math.Anal.*, **17** (1986), 945-960.

[FPR] A. Fasano, M. Primicerio, R. Ricci. Limiting behaviour of some problems in diffusive penetration. *Rendiconti di Mat.*, **VII-9** (1989), 39-57.

[FR] A. Fasano, R. Ricci. Penetration of solvents into glassy polymers, in *Free boundary problems: applications and theory* , A. Bossavit, A. Damlamian, M. Fremond ed.s, Research Notes in Math. **120**, Pitman, 1985.

[F1] A. Friedman. *Partial differential equation of parabolic type*, Prentice-Hall, 1964.

[F2] A. Friedman. *Variational principle and free-boundary problems*, Wiley, 1982.

[FB] G.F. Froment, K.B. Bischoff. *Chemical reactor design and analysis*, Wiley, New York, 1979.

[H] B. Hu. Diffusion of penetrantin a Polymer: a free boundary problem, preprint Univ. Minnesota, 1990.

[P] C.V. Pao. Asymptotic stability of reaction-diffusion systems in chemical reactor and combustion theory. *J.Math.Anal.Appl.*, **82** (1981), 503-526.

[SS] H.Y. Sohn, J. Szekely. A structural model for gas-solid reaction with a moving boundary. *Chem. Eng. Sci.*, **27** (1972), 763-778.

[SH] B. Sherman. Limiting behavior in some Stefan problems as the latent heat goes to zero. *SIAM J.Appl.Math.*, **20** (1971), 319-327.

[S] I. Stakgold. Reaction-diffusion problem in chemical engineering, in *Nonlinear diffusion problems*, A. Fasano, M. Primicerio ed.s, Lecture Notes in Math. 1224, Springer, 1986.

[SBG] I. Stakgold, K.B. Bischoff, V. Gokhale. Validity of the pseudo-steady-state approximation. *Int.J.Eng.Sci*, **21** (1983), 537-542.

[V] A. Visintin. Stafan problem with phase relaxation. *IMA J.Appl.Math.*, **34** (1985), 225-245.

RICCARDO RICCI

Dipartimento di Matematica "VITO VOLTERRA"
Universita di Ancona

# Micro-gravity

N D KAZARINOFF

# The role and effects of surface tension in convection in cylindrical float-zones

**Abstract.** In this paper the role and effects of surface tension in the development and continuation of oscillations in cylindrical float-zones is discussed. A physical mechanism is given for the continuation of asymmetric, axial oscillations of toroidal flow cells in such float-zones.

Consider a full, cylindrical float-zone (FZ) consisting of a low-Prandtl-number fluid such as liquid Si in a 0-g environment. Suppose that this FZ has a free, cylindrical surface S and flat ends held at the melting temperature $\theta_{Melt}$ of the liquid. Suppose that a ring-heater H surrounds the FZ above the mid-circle on S and that H radiates energy to the whole of S; see Fig. 1. Surface tension of the fluid forming S carries the energy provided by H and engenders convection in the FZ. In numerical simulations of such a full, cylindrical FZ, J. Wilkowski and the author [1-4] have found that as the temperature gradient $\Delta\theta \equiv \theta_{Heater} - \theta_{Melt}$ is increased beyond a critical value $\Delta\theta_c$, dependent upon the aspect ratio A = L/R of the FZ (L = length, R = radius), oscillations of increasing complexity occur in the low-Pr (Prandtl number) liquid and on S. Note that our model is axially symmetric; no azimuthal oscillations are allowed. Since our model is two-dimensional, the flows in the FZ have axial and radial components of velocity (u, v) in the axial and radial directions (x, r), respectively, where (x, r) are cylindrical coordinates.

Consider a cut K in the free surface S. Surface tension $\sigma$ is force/unit length in a direction tangent to S and perpendicular to K. V. G. Levich and V.S. Krylov have written [5] that "by virtue of surface tension it appears that the hydrodynamical regimes in a system with a mobile interface depend on the heat and mass transfer processes occuring in the system.

(A second excellent review article by S. H. Davis deals with thermocapillary instabilities [6], primarily in one and two-layer models, with heating from below)

Recently, I have found just how critical the role of surface tension $\sigma$ is in governing the flows in FZ's. The parameter plaing the key role in inergy input to the FZ through $\sigma$ is $\gamma \equiv |\Delta/\Delta\theta|$, the gradient of surface tension on S. For example, if (A) $\gamma$ is taken to be 0.43 dyn/cm/°K [7] and the $\sigma$ of Si at $\theta_{Melt}$ is $\sigma_0 = 720.0$ dyn./cm. [7], then at $\Delta\theta = 8K$ and $A = 0.75$ $(L = 1.0)$ the plot of axial velocity $u(0.5, t)$ on S at the midcircle of the FZ vs $u(0.25, t)$ on S (Fig. 2), resembles the plot (Fig. 3) of the same quantities for $\Delta\theta = 17K$ and $A = 0.75$ if I use the values (B) $\nabla\sigma$ = 0.28 dyn/cm/°K and $\sigma_0 = 870.0$ dyn/cm., supplied by D. Schwabe (private communication). This resemblance vanishes between the cases (A) and (B) as $\theta_H$ increases. For example, in case (A) if $\Delta\theta = 11K$ and $A = 0.75$ the plot of $u(0.5, t)$ on S vs $u(0.25, t)$ on S (Fig. 3) is quite different from the plot of the same quantities in case (B) for $\Delta\theta = 40K$ and $A = 0.75$ (Fig. 4). Fig. 4 much more closely resembles the plots of these quantities for $\Delta\theta = 13K$ and $A = 0.5$ (Fig. 5). Thus, lowering $\nabla\sigma$ at a $A = .75$ causes the development of the flows in the FZ to resemble that for the higher $\gamma$ at the lower $A = 0.5$, but the development occurs over a much wider range of $\Delta\theta$ at the lower values of $\gamma$.

To obtain the numerical results reported above the relation $\sigma = \sigma_0 - \gamma\Delta\theta$ is used. The surface tension enters the problem through the nonlinear boundry conditions on $S = \{(x, r) \mid r = h(x, t), h(0, t) = h(L, t) = R\}$ at time t, which are:

$$-(p - p_0) + (2\mu/N^2)[v_r + (h_x)2u_x - h_x(v_x + u_r)] = -2\kappa\sigma,$$
$$(\mu/N^2)[2h_x(v_r - u_x) - (v_x + u_r)(1 - h_x^2)] = -\sigma_x/N,$$

where $N(x,t) = [1 + h_x(x, t)^2]^{1/2}$, the mean curvature of $S(x, t)$ is

$$\kappa(x, t) = (1/2)[1/h - h_{xx}/N^{3/2}],$$

and we assume $p_0 = 0$. The quantity $\sigma_x$ appears in the second of the above b.c. 's on S and $\sigma_r$ is omitted there because the latter is small.

As can be seen from Figs. 4 and 5, the flow is asymmetric about the midplane of the FZ perpendicular to its axis. What causes this asymmetry, I

do not know. Perhaps it is machine round-off, or the order of treatment of the nodes in the finite-difference grid used, but it is an initially small effect. Each simulation is begun with $(u, v, \theta - \theta_{Melt}) = (0, 0, 0)$ so that initially there is symmetry with respect to the midplane $x = L/2$.

J. Wilkowski and I [4] find that the critical Marangoni number and the critical temperature gradient ($^\circ$K/cm) in their simulations are higher than those found in experiments reported by D. Schwabe, R. Velten and A. Scharmann [8]. They report $Ma_c \approx 100$ for $A = 1.0$; we obtain $Ma_c \approx 300$. Their experiments were conducted on half-zones heated from above and from below. Recognizing that this mode of heating differs from that used in industry for full FZ's, these authors write that "In real FZ growth (e.g. with RF heating, electron-beam heating or radiation heating), the heat is applied onto the free surface of the zone and diffuses and convects into the bulk liquid. Therefore, for thermocapillary forces, the growth situation is hydro-dynamically more stable than in our model situation. ..." [8, p.1263]. This may be a reason for the differences between their results and ours. For in our simulation heat is applied to all of S (in a normal distribution with the peak on the midcircle) and slowly diffuses and convects into the liquid Si.

The pictures of the flow during these oscillations are instructive; see Figs. 6 and 7, which are snapshots of the velocity field in the FZ at two different times t. The dots in these figures mean the vectors at these points were too long to plot. The two large toroidal flow-cells swing back and forth in the x-direction as a compound pendulum. Surface tension forces are responsible for this swing. The fluid near S is cooler at the ends, where $x = 0$ and $x = L$. Because of the increase in $\sigma$ toward $x = 0$ (one models $\sigma$ as a function of the temperature $\theta$ with $\sigma$ decreasing as $\theta$ increases) and because the fluid Si is viscous, flow occurs near to S away from the mid-plane $x = L/2$ toward the ends of the FZ. When the flow becomes asymmetric with respect to $x = L/2$, then oscillations begin because of the effects of $\sigma$. Namely, as fluid flows toward $x = L$, for example, it carries warmer fluid next to S and near to $x = L/2$ with it. Cooler fluid rises from the interior of the FZ near $x = L/4$ to replace it. The resulting decrease in $\theta$ on S toward $x = L/4$ causes the fluid to reverse its flow under the force exerted by $\sigma$. When the flow is sufficiently developed toward $x = 0$, the same phenomena occurs in reverse. Thus, asymmetric, nonsteady-state, oscillatory flow is produced. (The role of the tiny waves on S with amplitudes $\approx 10^{-4}$cm in this senario is unclear to me. However, were there no $\sigma$, the axial oscillations of the fluid cells could not take place. Indeed, they were not observed by D. Schwabe, *et al.* [8] (because they heat all of one flat end of their experimental half-FZ relative to the other end) and Rupp *et al.* [9] (because they do not admit a truly mobile free surface that feels the gradient in $\sigma$).

On the other hand, Jurisch and Löser have observed at least one toroidal cell in each half-FZ of a single-crystal, full FZ of Mo [10]. Jurisch's experiments on full FZ's of Mo and Nb yield $Ma_c$'s of 638 for A = 1.3, 925 for A = 1.1, and 1625 for A = 0.9, with oscillations of $\theta$ on S having amplitudes from $\approx 0.1$K to $\approx 1.25$K [11] as Ma is increased above $Ma_c$. Jurisch [11] does not report an S-shaped surface temperature profile as Kamotani *et al.* [12] do for fluids with Pr >> 1. The amplitudes of oscillations of the surface temperature found by Wilkowski and Kazarinoff [4] agree well with those Jurisch reports[11]. We also did not find an S-shaped surface temperature profile. Rupp et al report $Ma_c \approx 300$ for Si and periods of azimuthal oscillations of $\approx 6$s for Ma near to $Ma_c$.

## REFERENCES

1. N. D. Kazarinoff and J. S. Wilkowski, A numerical study of flows in zone-refined silicon crystals *Physics of Fluids A*, 1(1989), 625- 627.
2. N. D. Kazarinoff and J. S. Wilkowski, Marangoni flows in a cylindrical liquid bridge of silicon, pp. 65-73 in *Numerical Simulation of Oscillatory Convection in low-Pr fluids*, B. Roux, ed., vol. 27 of *Notes on Fluid Mechanics*, Vieweg, Braunschweig, 1990.

3. N. D. Kazarinoff and J. S. Wilkowski, Period tripling and subharmonic oscillations in Marangoni flows in a cylindrical liquid bridge, pp. 265-283 in *Asymptotic and computational analysis,* R. Wong, ed., Marcel Dekker, Inc., N. Y. and Basel, 1990.

4. N. D. Kazarinoff and J. S. Wilkowski, Bifurcations of numerically simulated flows in axially symmetric float-zones, *Physics of Fluids A,* to appear.

5. V. G. Levich and V.S. Krylov, Surface-tension-driven phenomena, *Ann. Rev. of Fluid Mech.,* **1**(1969), 293-316.

6. S. H. Davis, Thermocapillary instabilities, *Ann. Rev. of Fluid Mech.,* **19** (1987), 403-435.

7. N. Kobayashi, Computer simulation of the steady flow in a cylindrical floating zone under low gravity, *J. Crystal Growth,* **66**(1984), 63-72.

8. D. Schwabe, R. Velten and A. Scharmann, The instability of surface tension driven flows in models for floating zones under normal and reduced gravity, *J. Crystal Growth,* **99**(1990), 1258-1264.

9. R. Rupp, G. Müller, and G. Neumann, Three dimensional time dependent modelling of the Marangoni convection in zone melting configurations for GaAs, J. Crystal Growth, **97**(1989), 34-41.

10. M. Jurisch and W. Löser, Analysis of periodic non-rotational W striations in Mo single crystals due to nonsteady thermocapillary convection, *J. Crystal Growth,* **102**(1990), 214-222.

11. M. Jurisch, Surface temperature oscillations of a floating zone resulting from oscillatory thermocapillary convection, *J. Crystal Growth,* **102** (1990), 223-232.

12. Y. Kamotani, S. Ostrach and M. Vargas, Oscillatory thermocapillary convection in a simulated floating zone configuration, *J. Crystal Growth,* **66**(1984), 83-90.

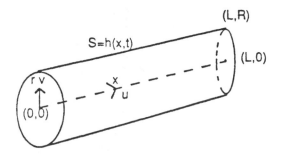

Figure 1

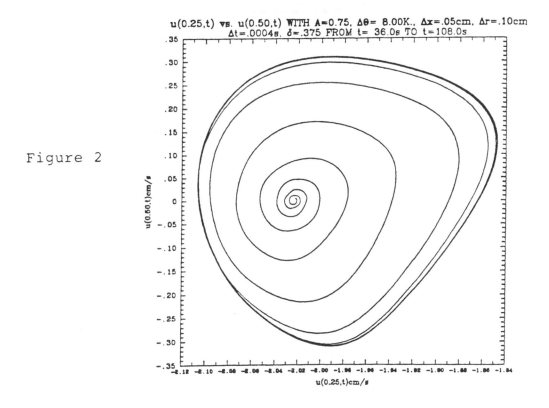

Figure 2

101

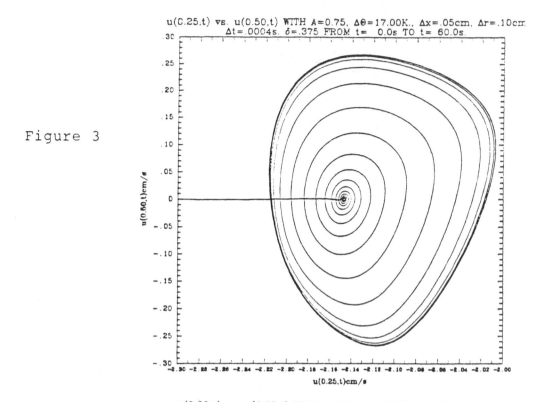

Figure 3

u(0.25,t) vs. u(0.50,t) WITH A=0.75, Δθ=17.00K., Δx=.05cm, Δr=.10cm.
Δt=.0004s. δ=.375 FROM t=  0.0s TO t= 60.0s.

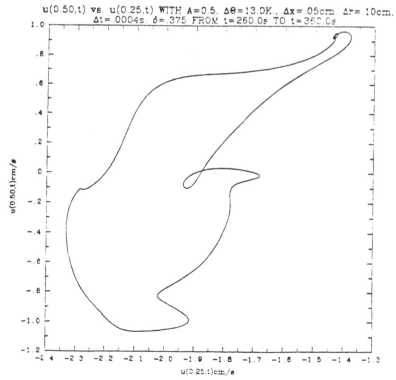

Figure 4

u(0.50,t) vs. u(0.25,t) WITH A=0.5, Δθ=13.0K., Δx=.05cm. Δr=.10cm.
Δt=.0004s. δ=.375 FROM t=260.0s TO t=360.0s

102

Figure 5

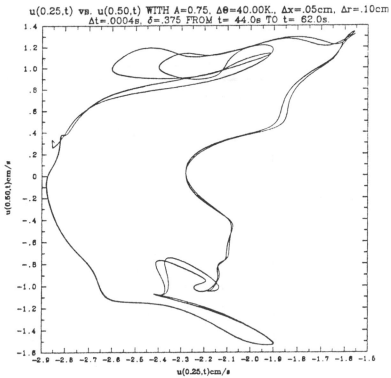

u(0.25,t) vs. u(0.50,t) WITH A=0.75, Δθ=40.00K., Δx=.05cm, Δr=.10cm
Δt=.0004s, δ=.375 FROM t= 44.0s TO t= 62.0s.

Figure 6

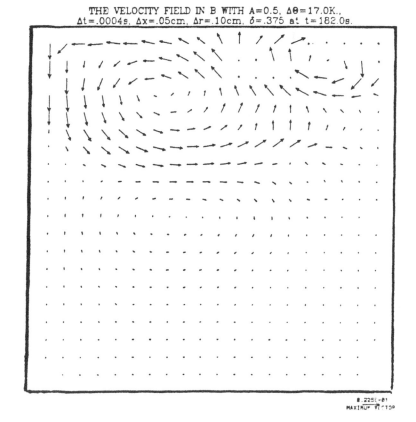

THE VELOCITY FIELD IN B WITH A=0.5, Δθ=17.0K.,
Δt=.0004s, Δx=.05cm, Δr=.10cm, δ=.375 at t=182.0s.

103

THE VELOCITY FIELD IN B WITH A=0.5, Δθ=17.0K.,
Δt=.0004s. Δx=.05cm. Δr=.10cm. δ=.375 at t=188.0s.

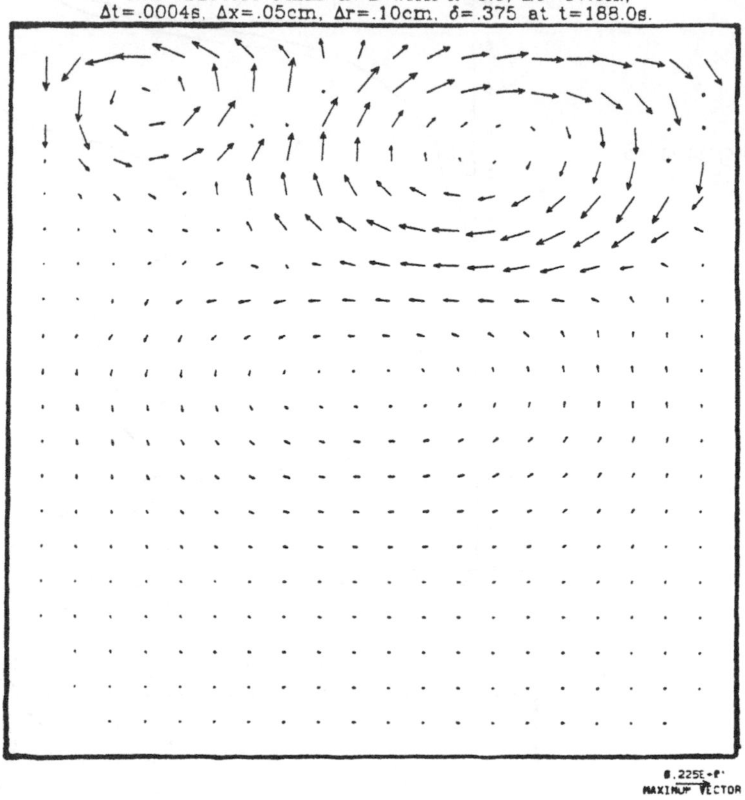

Figure 7

0.225E-8
MAXIMUM VECTOR

G B McFADDEN, B T MURRAY, S R CORIELL, M E GLICKSMAN AND
M E SELLECK

# Effect of a crystal-melt interface on Taylor-vortex flow with buoyancy

## 1 Introduction

The study of crystal growth from the liquid or melt phase provides a rich source of free boundary problems, the solutions for which are of great practical interest. (Some descriptions of common crystal growth techniques are given by Hurle and Jakeman in [1] and references therein.) When growing crystals of doped semiconductors or metallic alloys, the concentration of solute at the freezing interface is of special concern [2]. In most applications it is desirable to produce crystals with homogeneous distributions of solute throughout the crystal, and great care is taken in the design of the crystal growth apparatus to attempt to control the concentration and thermal fields near the interface. In directional solidification from the melt, for example, an idealized furnace that is free from imperfections would produce a planar crystal-melt interface with one-dimensional temperature and solute fields, allowing solute to be incorporated uniformly in the growing crystal. In reality, even such a planar crystal-melt interface is subject to various instabilities [3, 4] which can result in segregation of solute at the interface and produce inhomogeneous distributions of solute in the crystal. In addition to instabilities associated with the interface itself [5], under terrestrial growth conditions it is often difficult to avoid the occurrence of fluid flow in the melt due to natural convection [6]. Such flows are themselves able to cause undesirable segregation of solute [7] and may result in the production of inferior quality crystal. Avoiding natural convection is one of the main motivations for developing the capability of crystal growth under the microgravity conditions available in low earth orbit, where the driving force for natural convection is lower by orders of magnitude. In addition, such an environment allows more precise fundamental experiments on interface dynamics to be performed without the complicating effects of buoyancy-driven convection [8].

The study of the interaction of fluid flow with a crystal-melt interface is thus an area of fundamental importance in materials science, but despite much recent research [9] the understanding of such interactions is fragmentary. The general problem combines the complexities of the Navier-Stokes equations for the fluid flow in the melt with the nonlinear behavior of the free boundary representing the crystal-melt interface. Some progress has been made by studying explicit flows that allow a base state corresponding to a one-dimensional crystal-melt interface with solute and/or temperature fields that depend only on the distance from the interface. This allows the strength of the interaction between the flow and the interface to be assessed by a linear stability analysis of the simple base state.

For example, one can examine changes in the *morphological stability* [5] of the interface in the presence of flow in the melt. Specific flows that have been considered in this way include plane Couette flow [10, 11], thermosolutal convection [7, 12], plane stagnation flow [13, 14], rotating disk flow [15], and the asymptotic suction profile [16]. One can also examine changes in the *hydrodynamic stability* of a given flow that occur when a rigid bounding surface is replaced by a crystal-melt interface. Examples here include the instabilities associated with Rayleigh–Bénard convection [17], thermosolutal convection [6, 18, 19], plane Poiseuille flow [20], the asymptotic suction profile [16], thermally-driven flow in an annulus [20, 21], and

Taylor Couette flow [22, 23].

In previous work we have described the interaction of a Taylor-Couette flow with a cylindrical crystal-melt interface under the assumption that the effects of buoyancy can be neglected [22, 23]. This preliminary work is in support of experimental studies being conducted with succinonitrile; for this material the crystal-melt interface is predicted to have a significant effect on the conditions for marginal stability of the flow. Since large temperature differences in the system are capable of driving natural convection in the melt [24, 25], it is desirable to include such effects in our theoretical treatment as well. In this paper we reformulate the problem to include the effects of buoyancy when the axis of the cylinders is aligned with the direction of gravity. This generates a more complicated flow field in the base state, which then is subject to not only centrifugal instabilities but buoyant instabilities as well [26, 27]. We also include the effects of density-driven convection produced by the interaction between the radial density gradient and the centripetal acceleration of the azimuthal flow in the base state [28]. Linear stability results are obtained numerically for a typical experimental configuration. For these conditions we find that the effect of the contribution from the interaction of the radial density gradient with the centripetal acceleration is not significant. The natural convection causes a slight stabilization of the system; however, there is a two-fold increase in the wavelength of the most dangerous disturbance.

# 2  Taylor-Couette Flow with Buoyancy

We consider Taylor-Couette flow [29, 30] in the presence of a crystal-melt interface $\bar{r} = \bar{h}(\bar{z}, \phi, \bar{t})$ (overbars will denote dimensional quantities). A cylindrical coordinate system $(r, z, \phi)$ is used. The melt occupies the region $\bar{h}(\bar{z}, \phi, \bar{t}) < \bar{r} < \bar{R}_2$, and the crystal occupies the region $\bar{R}_0 < \bar{r} < \bar{h}(\bar{z}, \phi, \bar{t})$ (see Fig. 1); in the unperturbed base state the interface is an infinite cylinder with $\bar{h} = \bar{R}_1$. We also consider the convection-diffusion equation for heat transport as well. We consider steady rotation of the system; the outer cylinder is stationary and the inner cylinder and crystal rotate with angular velocity $\bar{\Omega}_1$. The temperature dependence of the density is taken into account in a Boussinesq approximation [31], which allows convective effects due to the interaction of the density gradient with the gravitational and centripetal acceleration terms. We assume the axis of the cylinders is aligned with the gravity vector, $-g\hat{z}$, where $\hat{z}$ is the unit vector along the $z$-axis.

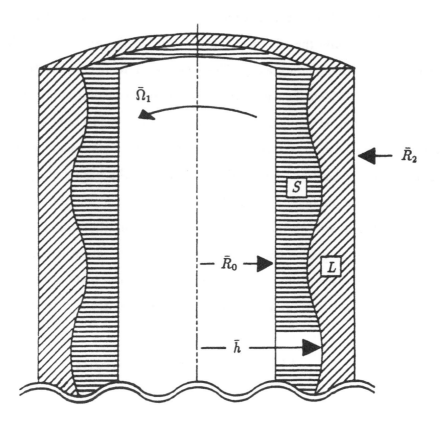

FIG. 1. Schematic diagram of the crystalline inner annulus (labeled "S") surrounded by the liquid phase (labeled "L"). In the base state the unperturbed crystal-melt interface is cylindrical, with $\bar{h}(\bar{z}, \phi, \bar{t}) = \bar{R}_1$.

The nonlinear dimensional governing equations in the melt are the continuity equation,

$$\frac{\partial \bar{u}}{\partial \bar{r}} + \frac{\bar{u}}{\bar{r}} + \frac{1}{\bar{r}}\frac{\partial \bar{v}}{\partial \phi} + \frac{\partial \bar{w}}{\partial \bar{z}} = 0, \tag{1a}$$

the momentum equations,

$$\bar{\rho}_0 \frac{D\bar{u}}{D\bar{t}} + \frac{\partial \bar{p}}{\partial \bar{r}} = \mu \left( \bar{\nabla}^2 \bar{u} - \frac{\bar{u}}{\bar{r}^2} - \frac{2}{\bar{r}^2}\frac{\partial \bar{v}}{\partial \phi} \right) + \bar{\rho}\frac{\bar{v}^2}{\bar{r}}, \tag{1b}$$

$$\bar{\rho}_0 \frac{D\bar{v}}{D\bar{t}} + \frac{1}{\bar{r}}\frac{\partial \bar{p}}{\partial \phi} = \mu \left( \bar{\nabla}^2 \bar{v} - \frac{\bar{v}}{\bar{r}^2} + \frac{2}{\bar{r}^2}\frac{\partial \bar{u}}{\partial \phi} \right) - \bar{\rho}\frac{\bar{u}\bar{v}}{\bar{r}}, \tag{1c}$$

$$\bar{\rho}_0 \frac{D\bar{w}}{D\bar{t}} + \frac{\partial \bar{p}}{\partial \bar{z}} = \mu \bar{\nabla}^2 \bar{w} - \bar{\rho}g, \tag{1d}$$

and the energy equation

$$\frac{D\bar{T}}{D\bar{t}} = \kappa \bar{\nabla}^2 \bar{T}, \tag{1e}$$

where

$$\frac{D}{D\bar{t}} = \frac{\partial}{\partial \bar{t}} + \bar{u}\frac{\partial}{\partial \bar{r}} + \frac{\bar{v}}{\bar{r}}\frac{\partial}{\partial \phi} + \bar{w}\frac{\partial}{\partial \bar{z}},$$

and

$$\bar{\nabla}^2 = \frac{\partial^2}{\partial \bar{r}^2} + \frac{1}{\bar{r}}\frac{\partial}{\partial \bar{r}} + \frac{1}{\bar{r}^2}\frac{\partial^2}{\partial \phi^2} + \frac{\partial^2}{\partial \bar{z}^2}.$$

Here $\bar{u}$, $\bar{v}$, and $\bar{w}$ are the velocity components in the $\bar{r}$, $\phi$, and $\bar{z}$ directions, respectively, $\bar{p}$ is the pressure, $\bar{T}$ is the temperature in the melt, $\mu$ is the viscosity coefficient, $\kappa$ is the thermal diffusivity in the melt, and $\bar{\rho} = \bar{\rho}_0(1 - \alpha[\bar{T} - \bar{T}_E])$ is the density, where $\bar{\rho}_0$ and $\bar{T}_E$ are reference densities and temperatures, and $\alpha$ is the coefficient of thermal expansion. In the Boussinesq approximation employed here [28, 31], the temperature variation in the density is neglected in the advection terms on the left hand sides of the momentum equations.

The solid is rotating with an azimuthal velocity $\bar{r}\bar{\Omega}_1$, and the temperature in the solid, $\bar{T}_S$, obeys

$$\frac{\partial \bar{T}_S}{\partial \bar{t}} + \bar{\Omega}_1 \frac{\partial \bar{T}_S}{\partial \phi} = \kappa_S \bar{\nabla}^2 \bar{T}_S, \tag{1f}$$

where $\kappa_S$ is the thermal diffusivity in the crystal.

At the outer boundary $\bar{r} = \bar{R}_2$, we have $\bar{u} = \bar{w} = \bar{v} = 0$, and $\bar{T} = \bar{T}_2$, where $\bar{T}_2$ is the (constant) temperature imposed at the outer cylinder; $\bar{T}_2$ is assumed to exceed the melting point $\bar{T}_m$ of the material. At the inner boundary $\bar{r} = \bar{R}_0$, we have $\bar{T}_S = \bar{T}_0$, where $\bar{T}_0$ is the (constant) temperature imposed at the inner cylinder; $\bar{T}_0$ is assumed to lie below $\bar{T}_m$. At the interface $\bar{r} = \bar{h}(\bar{z}, \phi, \bar{t})$ the boundary conditions are [6]

$$\bar{u} = \bar{w} = 0, \tag{2a}$$

$$\bar{v} = \bar{h}(\bar{z}, \phi, \bar{t})\bar{\Omega}_1, \tag{2b}$$

$$\bar{T} = \bar{T}_S = \bar{T}_m - \bar{T}_m \Gamma \bar{K}, \tag{2c}$$

$$-L_V \frac{\partial \bar{h}}{\partial \bar{t}} = k_L \left( \frac{\partial \bar{T}}{\partial \bar{r}} - \frac{1}{\bar{h}^2} \frac{\partial \bar{h}}{\partial \phi} \frac{\partial \bar{T}}{\partial \phi} - \frac{\partial \bar{h}}{\partial \bar{z}} \frac{\partial \bar{T}}{\partial \bar{z}} \right) - \tag{2d}$$

$$k_S \left( \frac{\partial \bar{T}_S}{\partial \bar{r}} - \frac{1}{\bar{h}^2} \frac{\partial \bar{h}}{\partial \phi} \frac{\partial \bar{T}_S}{\partial \phi} - \frac{\partial \bar{h}}{\partial \bar{z}} \frac{\partial \bar{T}_S}{\partial \bar{z}} \right),$$

where $\Gamma$ is a capillary length, $L_V$ is the latent heat of fusion per unit volume of crystal, $\bar{K}$ is the mean curvature of the interface, and $k_L$ and $k_S$ are the thermal conductivities in the liquid and solid, respectively. We have assumed equal densities of crystal and melt, and equal heat capacities in each phase.

## 3  Base State

The equations admit an annular base state with $\bar{h}(\bar{z}, \phi, \bar{t}) = \bar{R}_1$, $\bar{u} = 0$, $\bar{v} = \bar{V}(\bar{r})$, $\bar{w} = \bar{W}(\bar{r})$, $\bar{p} = \bar{P}(\bar{r}, \bar{z})$, $\bar{T} = \bar{\Theta}(\bar{r})$, and $\bar{T}_S = \bar{\Theta}_S(\bar{r})$. Dimensionless variables may be introduced as follows. The length scale is chosen to be the melt gap width $\bar{L} = \bar{R}_2 - \bar{R}_1$, the time scale is chosen to be $\bar{L}^2/\nu$, where $\nu = \mu/\bar{\rho}_0$ is the kinematic viscosity, the velocity scale is $\nu/\bar{L}$, the pressure scale is $\bar{\rho}_0\nu^2/\bar{L}^2$, and the deviation of the temperature from its value at the unperturbed interface is measured in units of the temperature difference across the melt, $\Delta T = \bar{T}_2 - \bar{T}_m + \bar{T}_m \Gamma / \bar{R}_1$. The melt then occupies the region $\eta/(1 - \eta) < r < 1/(1 - \eta)$, the unperturbed interface is located at $r = \eta/(1 - \eta)$, and the crystal occupies the region $\eta_S/(1 - \eta) < r < \eta/(1 - \eta)$, where $\eta = \bar{R}_1/\bar{R}_2$ and $\eta_S = \bar{R}_0/\bar{R}_2 < \eta$. (Dimensionless counterparts to the dimensional variables will lack overbars.)

The resulting dimensionless expressions for the base state variables are as follows. The base azimuthal velocity can be written in the form

$$V(r) = \frac{-\operatorname{Re}\eta^2}{(1 - \eta^2)}r + \frac{1}{r}\frac{\operatorname{Re}\eta^2}{(1 - \eta^2)(1 - \eta)^2},$$

where $\operatorname{Re} = \bar{L}^2\bar{\Omega}_1/\nu$ is the Reynolds number. The solution for the axial velocity that is appropriate to model a closed system with no net axial volume flux,

$$\int_{R_1}^{R_2} \bar{r}\,\bar{W}(\bar{r})\,d\bar{r} = 0, \tag{3}$$

is given by

$$W(r) = \frac{-G}{16(1 - \eta)^2} \left( C\left[\xi^2 - 1 + (1 - \eta^2)\frac{\ln\xi}{\ln\eta}\right] - 4(\xi^2 - \eta^2)\frac{\ln\xi}{\ln\eta} \right),$$

where $G = g\alpha\bar{L}^3\Delta T/\nu^2$ is the Grashof number, $\xi = (1 - \eta)r$, and

$$C = \frac{(1 - \eta^2)(1 - 3\eta^2) - 4\eta^4\ln\eta}{(1 - \eta^2)^2 + (1 - \eta^4)\ln\eta}.$$

The dimensionless temperature fields are given by

$$\Theta(r) = \frac{\ln(\xi/\eta)}{\ln(1/\eta)}, \tag{4a}$$

and

$$\Theta_S(r) = \left(\frac{k_L}{k_S}\right)\frac{\ln(\xi/\eta)}{\ln(1/\eta)}, \tag{4b}$$

The radial pressure gradient balances the centrifugal force, and the axial pressure gradient is constant.

# 4 Linearized equations

We next consider the dimensionless equations linearized about the steady base state. We Fourier transform the axial and azimuthal coordinates, which introduces the axial wavenumber $a$ and the azimuthal wavenumber $n$, and write

$$
\begin{pmatrix} u(r,z,\phi,t) \\ v(r,z,\phi,t) \\ w(r,z,\phi,t) \\ p(r,z,\phi,t) \\ T(r,z,\phi,t) \\ T_S(r,z,\phi,t) \\ h(z,\phi,t) \end{pmatrix} = \begin{pmatrix} 0 \\ V(r) \\ W(r) \\ P(r,z) \\ \Theta(r) \\ \Theta_S(r) \\ \eta/(1-\eta) \end{pmatrix} + \begin{pmatrix} \hat{u}(r) \\ \hat{v}(r) \\ \hat{w}(r) \\ \hat{p}(r) \\ \hat{\Theta}(r) \\ \hat{\Theta}_S(r) \\ \hat{h} \end{pmatrix} \exp(\sigma t + in\phi + iaz),
$$

where the perturbation amplitudes (quantities with hats) are assumed small. The complex growth rate $\sigma = \sigma_r + i\sigma_i$ determines marginal stability: the flow is stable if $\sigma_r < 0$ for all values of $n$ and $a$. If one sets $D = \partial/\partial r$, the linearized equations in the melt region $\eta/(1-\eta) < r < 1/(1-\eta)$ take the form

$$
D\hat{u} + \frac{1}{r}\hat{u} + \frac{in}{r}\hat{v} + ia\hat{w} = 0, \tag{5a}
$$

$$
\sigma\hat{u} + \frac{inV}{r}\hat{u} + iaW\hat{u} + D\hat{p} = \tag{5b}
$$

$$
\left(D^2\hat{u} + \frac{1}{r}D\hat{u} - \left[a^2 + \frac{(1+n^2)}{r^2}\right]\hat{u} - \frac{2in}{r^2}\hat{v}\right) + 2\frac{(1-\epsilon\Theta)V}{r}\hat{v} - \frac{\epsilon V^2}{r}\hat{\Theta},
$$

$$
\sigma\hat{v} + \frac{inV}{r}\hat{v} + iaW\hat{v} + \hat{u}DV + \frac{in}{r}\hat{p} = \tag{5c}
$$

$$
\left(D^2\hat{v} + \frac{1}{r}D\hat{v} - \left[a^2 + \frac{(1+n^2)}{r^2}\right]\hat{v} + \frac{2in}{r^2}\hat{u}\right) - \frac{(1-\epsilon\Theta)V}{r}\hat{u},
$$

$$
\sigma\hat{w} + \frac{inV}{r}\hat{w} + iaW\hat{w} + \hat{u}DW + ia\hat{p} = \tag{5d}
$$

$$
\left(D^2\hat{w} + \frac{1}{r}D\hat{w} - \left[a^2 + \frac{n^2}{r^2}\right]\hat{w}\right) + G\hat{\Theta},
$$

112

$$\sigma \hat{\Theta} + \frac{inV}{r}\hat{\Theta} + iaW\hat{\Theta} + \hat{u}D\Theta = \frac{1}{P_r}\left(D^2\hat{\Theta} + \frac{1}{r}D\hat{\Theta} - \left[a^2 + \frac{n^2}{r^2}\right]\hat{\Theta}\right), \qquad (5e)$$

and, in the region $\eta_S/(1-\eta) < r < \eta/(1-\eta)$, one obtains

$$\sigma \hat{\Theta}_S + in\mathrm{Re}\hat{\Theta}_S = \frac{1}{P_s}\left(D^2\hat{\Theta}_S + \frac{1}{r}D\hat{\Theta}_S - \left[a^2 + \frac{n^2}{r^2}\right]\hat{\Theta}_S\right). \qquad (5f)$$

Here $P_r = \nu/\kappa_L$ is the Prandtl number, $P_s = \nu/\kappa_S$, and $\epsilon = \alpha\Delta T$.

The linearized boundary conditions are $\hat{u} = \hat{v} = \hat{w} = \hat{\Theta} = 0$ at $r = 1/(1-\eta)$, $\hat{\Theta}_S = 0$ at $r = \eta_S/(1-\eta)$, and, at the interface $r = R_1 = \eta/(1-\eta)$,

$$\hat{u} = 0, \qquad (5g)$$

$$\hat{w} + \hat{h}DW = 0, \qquad (5h)$$

$$\hat{v} + \hat{h}DV = \mathrm{Re}\hat{h}, \qquad (5i)$$

$$\hat{\Theta} + \hat{h}D\Theta = \hat{\Theta}_S + \hat{h}D\Theta_S = -\gamma\left(a^2 + \frac{(n^2-1)}{R_1^2}\right)\hat{h}, \qquad (5j)$$

$$-\sigma\mathcal{L}\hat{h} = \left(D\hat{\Theta} - qD\hat{\Theta}_S\right). \qquad (5k)$$

where $\gamma = (\bar{T}_m\Gamma)/(\bar{L}\Delta T)$, $\mathcal{L} = (\nu L_V)/(k_L\Delta T)$, and $q = k_S/k_L$.

# 5    Numerical Results

The linearized equations constitute an eigenvalue problem from which the complex growth rate $\sigma = \sigma_r + i\sigma_i$ can be determined for each choice of the remaining parameters. Curves of marginal stability may be obtained by setting $\sigma_r = 0$ and computing instead a pair $(\mathrm{Re}, \sigma_i)$ as a function of wavenumber, which can be done (see [20]) in the manner suggested by Keller [32]. The numerical procedure was checked carefully against previously published linear stability results for related problems [22, 27, 28, 33].

To illustrate the relative importance of the effects of buoyancy on the stability of the system, we use the material properties of succinonitrile [21] for a typical container geometry. We take $\bar{R}_2 = 1.60$ cm, $\bar{R}_1 = 1.11$ cm, and $\bar{R}_0 = 0.458$ cm, giving $\eta = 0.690$ and $\eta_S = 0.286$. We assume a temperature gradient in the melt of 2.5 K/cm, which produces a temperature difference across the melt gap of $\Delta T = 1.225$ K, giving $\mathcal{L} = 455$, $\epsilon = 9.923 \cdot 10^{-4}$, $G = 170$, and $\gamma = 4.9 \cdot 10^{-3}$; for simplicity we set $\gamma = 0$. The thermal properties of the liquid and solid are similar for succinonitrile, and we take the ratio of thermal conductivities to be $q = 1$, with $P_r = P_s = 22.8$. We restrict our discussion to axisymmetric disturbances ($n = 0$), which are expected to be the most dangerous modes in this configuration.

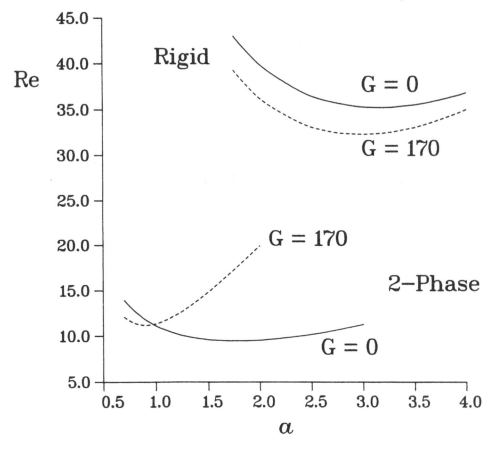

FIG. 2. Marginal values of the Reynolds number, Re, versus the axial wavenumber of the perturbation, $a$, for axisymmetric disturbances, comparing the effects of buoyancy ($G = 170$) for the rigid-walled system (top two curves) and for the system with a crystalline inner cylinder (lower two curves).

Marginal stability curves ($\sigma_r = 0$) are compared in Fig. 2 for four different cases. The Reynolds number Re of the flow is plotted versus the axial wavenumber $a$ of the disturbance. The curves in this plot are computed with $\epsilon = 0$, so that the interaction of the density gradient with the centripetal acceleration is neglected.

The top two curves correspond to the classical Taylor-Couette problem with rigid isothermal surfaces bounding the liquid, and the bottom two curves correspond to the two-phase problem with a crystal-melt interface at $\bar{r} = \bar{R}_1$. We first consider the two cases without buoyancy ($G = 0$). The rigid-walled system has a critical Reynolds number Re = 35 (see Table I), corresponding to the onset of the classical secondary flow consisting of toroidal Taylor-vortex cells. The crystal-melt interface destabilizes the system, giving a smaller critical Reynolds number Re = 9; the axial wavelength of the most dangerous disturbance, $\lambda = 2\pi/a$, is shifted to larger values. This effect has been described in our previous work [22, 23]; the size of the destabilization is found to vary with the Prandtl number of the melt. The results for a rigid-walled system actually provide the limiting values for a conduction-dominated system with $P_r \ll 1$. The destabilization due to a crystal-melt interface becomes pronounced for convection-dominated systems with $P_r \gg 1$. For $G = 0$ the disturbances are stationary in both cases, with $\sigma_i = 0$. For $G = 170$ the marginal curves are shifted, as indicated by the dashed curves in Fig. 2. The rigid-walled system is destabilized slightly from Re = 35 to Re = 32, with a small increase in wavelength. This destabilization of the Taylor-Couette flow by the buoyancy forces is consistent with the trends reported previously [27]. The toroidal cells are no longer stationary, but drift downwards with an axial phase velocity equal to $-\sigma_i/a$. The effect of buoyancy when a crystal-melt interface is present is quite different, however. There is a slight increase in the critical Reynolds number when buoyancy is present, and there is a two-fold increase in the wavelength of the most dangerous disturbance. The secondary cells have a slow *upward* drift ($\sigma_i < 0$), but the phase velocity is smaller by three orders of magnitude. This decrease in phase velocity is likely due to the strong coupling between the interface and the flow: the interface deforms and accommodates the cellular flow structure, allowing the instability to occur at lower Reynolds numbers. For the cellular structure to translate in the axial direction, however, the deformed interface must melt and freeze for the wave to propagate, which retards the dynamics of the process.

## Table I

| $P_r$ | $G$ | $\epsilon$ | Re | $\sigma_i$ | $a$ |
|-------|-----|------------|----------|------------------------|-------|
| 0     | 0   | 0          | 35.2520  | 0                      | 3.143 |
| 0     | 0   | 0.001      | 35.3777  | 0                      | 3.143 |
| 0     | 170 | 0          | 32.3174  | 4.6937                 | 2.967 |
| 0     | 170 | 0.001      | 32.2991  | 4.6881                 | 2.967 |
| 22.8  | 0   | 0          | 9.4518   | 0                      | 1.782 |
| 22.8  | 0   | 0.001      | 9.4508   | 0                      | 1.782 |
| 22.8  | 170 | 0.0        | 11.1517  | $-2.7096 \cdot 10^{-3}$ | 0.904 |
| 22.8  | 170 | 0.001      | 11.1509  | $-2.7101 \cdot 10^{-3}$ | 0.904 |

Critical values of the Reynolds number Re, the time constant $\sigma_i$, and the critical wavenumber $a$ for the rigid-walled case ($P_r = 0$) and the case of a crystalline inner annulus of succinonitrile ($P_r = 22.8$). Buoyancy forces are absent for $G = 0$, and the interaction of the centripetal acceleration with the radial density is absent for $\epsilon = 0$.

In Table I we give values of the critical Reynolds numbers, the time constant $\sigma_i$, and the critical wavenumber $a$ for the cases shown in Fig. 2, and we also give results for the same cases computed with a non-zero contribution from the interaction of the centripetal acceleration with the radial density gradient, using the value $\epsilon = 0.001$. The changes brought about by including density variation in the centripetal acceleration are seen to be insignificant for these cases. The larger change in the stability of the system with a radial temperature gradient is clearly due to gravitational forces.

# 6 Acknowledgements

This work was conducted with the support of the Microgravity Science and Applications Division of the National Aeronautics and Space Administration, and the Applied and Computational Mathematics Program of the Defense Advanced Research Projects Agency. One of the authors (BTM) was supported by an NRC Postdoctoral Research Fellowship.

# References

[1] D. T. J. Hurle and E. Jakeman, *Introduction to the techniques of crystal growth*, PCH PhysicoChemical Hydrodynamics **2** (1981) pp. 237–244.

[2] R. A. Brown, *Theory of transport processes in single crystal growth from the melt*, AIChE J. **34** (1988) pp. 881-911.

[3] S. R. Coriell, G. B. McFadden and R. F. Sekerka, *Cellular growth during directional solidification*, Ann. Rev. Mater. Sci. **15** (1985) pp. 119–145.

[4] M. E. Glicksman, S. R. Coriell and G. B. McFadden, *Interaction of flows with the crystal-melt interface*, Ann. Rev. Fluid Mech. **18** (1986) pp. 307-335.

[5] W. W. Mullins and R. F. Sekerka, *Stability of a planar interface during solidification of a dilute binary alloy*, J. Appl. Phys. **35** (1964) pp. 444-451.

[6] S. R. Coriell and R. F. Sekerka, *Effect of convective flow on morphological stability*, PCH PhysicoChem. Hydrodyn. **2** (1981) pp. 281–293.

[7] S. R. Coriell, M. R. Cordes, W. J. Boettinger, and R. F. Sekerka, *Convective and interfacial instabilities during unidirectional solidification of a binary alloy*, J. Crystal Growth **49** (1980) pp. 13-28.

[8] M. E. Glicksman, E. Winsa, R. C. Hahn, T. A. Lograsso, S. H. Tirmizi, and M. E. Selleck, *Isothermal dendritic growth – a proposed microgravity experiment*, Metall. Trans. **19A** (1988) pp. 1945–1953.

[9] S. H. Davis, *Hydrodynamic interactions in directional solidification*, J. Fluid. Mech. **212** (1990) pp. 241–262.

[10] R. T. Delves, *Theory of Interface Stability*, in *Crystal Growth*, B. R. Pamplin, ed., Pergamon, Oxford, 1974, pp. 40-103.

117

[11] S. R. Coriell, G. B. McFadden, R. F. Boisvert, and R. F. Sekerka, *Effect of a forced Couette flow on coupled convective and morphological instabilities during unidirectional solidification*, J. Crystal Growth **69** (1984) pp. 15-22.

[12] S. R. Coriell and G. B. McFadden, *Buoyancy effects on morphological instability during directional solidification*, J. Crystal Growth **94** (1989) pp. 513-521.

[13] K. Brattkus, and S. H. Davis, *Flow induced morphological instability: stagnation point flows*, J. Crystal Growth **89** (1988) pp. 423-427.

[14] G. B. McFadden, S. R. Coriell, and J. I. D. Alexander, *Hydrodynamic and free boundary instabilities during crystal growth: the effect of a plane stagnation flow*, Comm. Pure and Appl. Math **41** (1988) pp. 683-706.

[15] K. Brattkus and S. H. Davis, *Flow induced morphological instability: the rotating disk*, J. Crystal Growth **87** (1988) pp. 385-396.

[16] S. A. Forth and A. A. Wheeler, *Hydrodynamic and morphological stability of the unidirectional solidification of a freezing binary alloy: a simple model*, J. Fluid Mech. **202** (1989) pp. 339-366.

[17] S. H. Davis, U. Müller, and C. Dietsche, *Pattern selection in single-component systems coupling Bénard convection and solidification*, J. Fluid. Mech. **144** (1984) pp. 133–151.

[18] B. Caroli, C. Caroli, C. Misbah, and B. Roulet, *Solutal convection and morphological instability in directional solidification of binary alloys*, J. Phys. (Paris) **46** (1985) pp. 401-413.

[19] G. W. Young and S. H. Davis, *Directional solidification with buoyancy in systems with small segregation coefficient*, Phys. Rev. **B34** (1986) pp. 3388-3396.

[20] G. B. McFadden, S. R. Coriell, R. F. Boisvert, M. E. Glicksman, and Q. T. Fang, *Morphological stability in the presence of fluid flow in the melt*, Metall. Trans. **15A** (1984) pp. 2117-2124.

[21] Q. T. Fang, M. E. Glicksman, S. R. Coriell, G. B. McFadden, and R. F. Boisvert, *Convective influence on the stability of a crystal-melt interface*, J. Fluid Mech. **151** (1985) pp. 121-140.

[22] G. B. McFadden, S. R. Coriell, M. E. Glicksman, and M. E. Selleck, *Instability of a Taylor-Couette flow interacting with a crystal-melt interface*, PhysicoChem. Hydrodyn. **11** (1989) pp. 387-409.

[23] G. B. McFadden, S. R. Coriell, B. T. Murray, M. E. Glicksman, and M. E. Selleck, *Effect of a crystal-melt interface on Taylor-vortex flow*, Phys. Fluids A **2** (1990) pp. 700-705.

[24] S. K. F. Karlsson and H. A. Snyder, *Observations on a thermally induced instability between rotating cylinders*, Ann. Phys. **31** (1965) pp. 314–324.

[25] K. S. Ball, B. Farouk, and V. C. Dixit, *An experimental study of heat transfer in a vertical annulus with a rotating inner cylinder*, Int. J. Heat Mass Transfer **32** (1989) pp. 1517–1527.

[26] K. S. Ball and B. Farouk, *Bifurcation phenomena in Taylor-Couette flow with buoyancy effects*, J. Fluid Mech. **197** (1988) pp. 479–501.

[27] M. E. Ali, The stability of Taylor-Couette flow with radial heating. Ph. D. Dissertation, Dept. of Mech. Eng., University of Colorado, Boulder, Colorado, 1988.

[28] J. Walowit, S. Tsao, and R. C. DiPrima, *Stability of flow between arbitrarily spaced concentric cylindrical surfaces including the effect of a radial temperature gradient*, Trans. Am. Soc. Mech. Eng. E: J. Appl. Mech. **31** (1964) pp. 585–592.

[29] G. I. Taylor, Phil. Trans. Roy. Soc. A **223**, 289 (1923).

[30] P. G. Drazin and W. H. Reid, *Hydrodynamic Stability*, Cambridge University Press, 1981.

[31] S. Chandrasekhar, *Hydrodynamic and Hydromagnetic Stability*, Oxford University Press, 1961.

[32] H. B. Keller, *Numerical Solution of Two Point Boundary Value Problems*, Regional Conference Series in Applied Mathematics **24**, SIAM, Philadelphia, 1976.

[33] G. B. McFadden, S. R. Coriell, R. F. Boisvert, and M. E. Glicksman, *Asymmetric instabilities in buoyancy-driven flow in a tall vertical annulus*, Phys. Fluids **27** (1984) pp. 1359–1361.

G. B. McFadden, B. T. Murray, and S. R. Coriell
National Institute of Standards and Technology
Gaithersburg, MD 20899

M. E. Glicksman and M. E. Selleck
Department of Materials Engineering
Rensselaer Polytechnic Institute    Troy, NY 12181

M K SMITH

# Linear and nonlinear stability of thermocapillary flows in microgravity

## Abstract

Thermocapillary flows can be a dominant source of motion in some material processing applications in microgravity environments. This paper describes how a simple liquid-layer model has been used to explore the various modes of instability in such flows.

## 1 Introduction

The float-zone method has been proposed as a means of growing high-quality single crystals in a microgravity environment. In this application, thermocapillary forces can drive vigorous fluid motions in the melt. If a steady thermocapillary flow in the zone becomes unstable, the quality of the final crystalline product may seriously degrade. This provides the motivation for a study of the instability of thermocapillary flows.

Relevant earth-based experiments on small model float zones have been done by Chun (1980), Preisser, Schwabe, & Scharmann (1983), and Kamotani, Ostrach, & Vargas (1984). Axisymmetric and non-axisymmetric modes of oscillation of the basic thermocapillary flow have been observed. Some experimental work seems to indicate that surface deformation is necessary for the instability.

Numerical work on model float zones or on two-dimensional cavities has been done by Kazarinoff & Wilkowski (1989), Rupp, Müller, & Neumann (1989), Shen, Neitzel, Jankowski, & Mittelmann (1990), and Carpenter & Homsy (1990). These studies suggest that surface deformation is a necessary part of an axisymmetric or two-dimensional thermocapillary instability, but that surface deformation is not needed when the unstable motion is non-axisymmetric or three dimensional.

The complexity of the flow geometry associated with the float-zone method prevents a direct theoretical description of the thermocapillary instability in this system. However, a simple model of a single liquid layer driven by thermocapillarity has been used to explore the mechanisms associated with various modes of instability. This work provides some insight into the behaviour of thermocapillary flows seen in a real float zone.

## 2  Mathematical Model

The model used by Smith & Davis (1983a,b), hereafter referred to as SD, is a single liquid layer of depth $d$ bounded below by a rigid, insulating plane (see Figure 1).

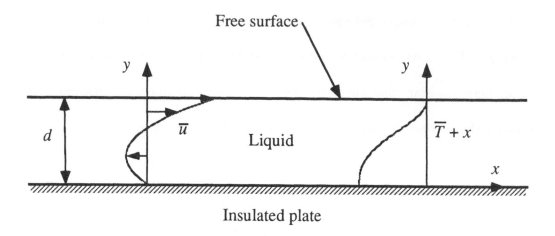

Figure 1: A schematic of the liquid-layer geometry. Also shown are the basic-state velocity and temperature profiles.

The upper surface of the liquid is free and bounded by a passive gas. Gravity is zero. The liquid is Newtonian with constant dynamic viscosity $\mu$, density $\rho$, kinematic viscosity $\nu$, thermal conductivity $k$, thermal diffusivity $\kappa$, and unit thermal surface conductance $h$. The surface tension $\sigma$ of the interface has an approximate equation of state $\sigma = \sigma_0 - \gamma(T - T_0)$, where $\sigma_0$ and $T_0$ are a reference surface tension and temperature respectively, and $\gamma$ is the negative of the rate of change of surface tension with temperature. The temperature of the passive gas varies linearly with $x$ so that a constant temperature gradient $dT/dx = -b$ is imposed along the interface.

The dimensionless groups that appear after the scaling of the governing equations are the Marangoni number $M = \gamma b d^2 / \mu \kappa$, the Prandtl number $Pr = \nu / \kappa$, the Biot number $Bi = hd/k$, and the surface-tension number $S = \rho d \sigma_0 / \mu^2$. The Reynolds number is $R = M Pr^{-1}$.

The dimensionless governing equations for this model are

$$R\{\underline{v}_t + (\underline{v} \cdot \nabla)\underline{v}\} = -\nabla p + \nabla^2 \underline{v} \tag{2.1a}$$

$$M\{T_t + \underline{v} \cdot \nabla T\} = \nabla^2 T, \quad \nabla \cdot \underline{v} = 0. \tag{2.1b,c}$$

Boundary conditions on the rigid plane at $z = 0$ are

$$\underline{v} = T_z = 0, \tag{2.1d,e}$$

and on the interface at $z = \eta(x, y, t)$ are

$$\eta_t + \underline{v} \cdot \nabla \eta = w, \quad \underline{n} \cdot \underline{\underline{\Sigma}} \cdot \underline{n} = 2H(SR^{-1} - T) \tag{2.1f,g}$$

$$\underline{t}^{(\beta)} \cdot \underline{\underline{\Sigma}} \cdot \underline{n} = -\underline{t}^{(\beta)} \cdot \nabla T, \quad \text{for } \beta = 1, 2 \tag{2.1h}$$

$$-\underline{n} \cdot \nabla T = Bi(T - T_\infty) + Q. \tag{2.1i}$$

Here, $\underline{n}$ and $\underline{t}^{(\beta)}$, $\beta = 1, 2$ are the unit normal and tangent vectors to the interface, $\underline{\underline{\Sigma}}$ is the stress tensor of the liquid, $2H$ is twice the mean curvature of the interface, $T_\infty$ is the temperature of the passive gas, and $Q$ is an imposed heat flux from the liquid to the gas.

If the surface tension of the interface is large enough so that the interface remains planar, a parallel shear flow driven by thermocapillarity can be found as shown in Figure 1. This flow corresponds to the core flow in a shallow slot found by Sen & Davis (1982), and it is the basic state of most direct interest for crystal-growth applications.

## 3    Linear Stability Analysis

SDa,b did a standard linear stability analysis of the basic-state flow shown in Figure 1. In SDa, they described an instability, called a hydrothermal wave, that is three dimensional and exists when the free surface is nondeformable. Because of the symmetry of the system, the instability takes the form of two planar waves; one moves upstream and to the right, while the other moves upstream and to the left. For small $Pr$, the critical Marangoni number $M_c$ decreases as $Pr$ decreases and the disturbance propagates almost transversely to the basic-state flow.

The mechanism for this thermocapillary instability has been described in detail by Smith (1986). For small Prandtl numbers, consider a hot thermal disturbance to the interface in the form of a line parallel to the $x$-axis. Thermocapillarity produces a transverse flow and a corresponding vertical flow that brings slow-moving fluid from the interior of the layer toward the interface. This upflow reduces the velocity of the streamwise interfacial flow, and since this flow is convectively heating the

124

fluid in the top portion of the layer, the temperature of the disturbance decreases. When the thermal disturbance disappears, the interfacial velocity is still less than its basic-state value and so the temperature of the interface continues to fall. The thermal disturbance becomes cold, thermocapillarity reverses the vertical flow, and the streamwise interfacial flow increases back to its basic-state value while the temperature of the disturbance attains a minimum. The same process now works in reverse and the hot disturbance is reformed. If the imposed interfacial temperature gradient is large enough, the energy input to the disturbance by streamwise convection overcomes the energy loss due to conduction and the temperature of the disturbance increases with each cycle. Thus, the process is unstable.

A second mode of instability, discussed by SDb, takes the form of a surface wave. To describe this mode, surface deformation is allowed and two-dimensional forms of the interfacial boundary conditions are used. The critical mode of instability is a long-wave mode whose wavenumber $\alpha = 0$. Using the regular perturbation method, SDb found the critical Reynolds number

$$R_c = \left\{ \frac{1}{3} \tilde{S} \left( \frac{1}{10} + \frac{1}{4} Pr Bi^{-1} \right)^{-1} \right\}^{1/2}, \tag{3.1}$$

where the parameter $\tilde{S} = \alpha^2 S = O(1)$ is used to bring in surface-tension effects at this order.

This mode of instability is directly related to the long-wave interfacial instability seen in inclined liquid layers. The physical mechanism can be explained using the arguments of Smith (1990). Consider an interfacial deformation whose wavelength is much larger than the depth of the layer. The resulting thermocapillary flow in the layer will tend to follow the fully-developed parallel flow dictated by the layer's depth. However, inertial effects in the fluid cause a phase shift between the interfacial deformation and this longitudinal flow. The result is a disturbance flow toward the crest of the disturbance that increases the interfacial deformation. Also, at the disturbance crest the temperature is reduced, producing a disturbance thermocapillary flow toward the crest that increases the interfacial deformation. Surface tension increases the pressure under a disturbance crest, creating a flow away from the crest that reduces the interfacial deformation. If the inertial and thermocapillary effects are larger than the surface-tension effect, the layer will be unstable.

In the slot geometry of Sen & Davis (1982), a long-wavelength disturbance will interact with the ends of the slot. SDb estimated this effect by doing finite wavenumber calculations and choosing the wavelength of the most dangerous disturbance to be equal to the slot length. Disturbances with longer wavelengths are assumed to be damped out by the ends. This estimate reveals that the surface-wave instability is preferred over the hydrothermal-wave instability for small enough Prandtl numbers.

# 4  Nonlinear Stability Theory

A nonlinear theory must be used to determine the post-critical behaviour of the three-dimensional hydrothermal waves examined by SDa. Smith (1988) used the multiple-scale method with the small parameter $\epsilon$, defined as $\epsilon^2 = (M - M_c)/M_c$, to derive nonlinear evolution equations for the amplitudes of the right- and left-moving waves when the system is just above the critical point from linear theory. The evolution equations in unscaled form are

$$A_t + \underline{c}_g^{(+)} \cdot \nabla A = L^{(+)}A + \epsilon^2 cA + A\left\{c_{aa}|A|^2 + c_{bb}|B|^2 + \underline{c}_p^{(+)} \cdot \nabla P\right\} \qquad (4.1a)$$

$$B_t + \underline{c}_g^{(-)} \cdot \nabla B = L^{(-)}B + \epsilon^2 cB + B\left\{c_{aa}|B|^2 + c_{bb}|A|^2 + \underline{c}_p^{(-)} \cdot \nabla P\right\} \qquad (4.1b)$$

$$0 = \frac{1}{3}\nabla^2 P + \underline{c}_{rp}^{(+)} \cdot \nabla|A|^2 + \underline{c}_{rp}^{(-)} \cdot \nabla|B|^2. \qquad (4.1c)$$

The following heat-flux condition augments these equations when $Bi = 0$,

$$Q = \frac{1}{3}M_c\overline{T}_x P_x + M_c c_q(|A|^2 + |B|^2). \qquad (4.2)$$

In these equations, $A$ and $B$ are the small amplitudes of the right- and left-moving hydrothermal waves and $P$ is the small amplitude of a pressure field induced by the hydrothermal waves. The operator $L^{(\pm)}$ is defined as

$$L^{(\pm)} = c_{xx}\frac{\partial^2}{\partial x^2} \pm c_{xy}\frac{\partial^2}{\partial x \partial y} + c_{yy}\frac{\partial^2}{\partial y^2}, \qquad (4.3)$$

and $\nabla$ and $\nabla^2$ are the gradient and the Laplacian operators in $x$ and $y$. The scalar and vector $c$'s with various subscripts and superscripts are complex or real constants obtained from orthogonality conditions.

Equilibrium solutions of these nonlinear equations correspond to pure $A$ or $B$ waves or to a mixture of both waves. When the stability of each of these solutions is examined, Smith (1988) found that pure waves are stable to amplitude disturbances for all values of $Bi$ and $Pr$ that he considered except for $Bi = 1$ and $Pr < 0.01$. In this small Prandtl number range, only mixed waves are stable to amplitude disturbances.

In addition, both pure and mixed waves have a sideband instability for all parameters considered. This sideband instability is a generalization of the one found in the one-dimensional complex Ginzburg-Landau equation and identified by the criterion of Newell (1974).

## 5   Conclusions

We have shown how a simple liquid-layer model can be used to examine the instability mechanisms of thermocapillary flows. The hydrothermal-wave instability is characterized by a three-dimensional motion and no deformation of the interface of the layer. For the surface-wave instability, the interface does deform and the motion is only two dimensional. These observations seem to agree with current computational results for the unstable motion in realistic float-zone geometries.

The research described in this paper was supported by the NASA Material-Processing-in-Space Program with a grant to Prof. S. H. Davis, and by the NSF Fluid Mechanics Program. The author thanks Prof. Davis for his support and guidance.

## REFERENCES

Carpenter, B. M. & Homsy, G. M. 1990 *Phys. Fluids A* **2**, 137.

Chun, C.-H. 1980 *Acta Astronautica* **7**, 479.

Kamotani, Y., Ostrach, S., & Vargas, M. 1984 *J. Crystal Growth* **66**, 83.

Kazarinoff, N. D. & Wilkowski, J. S. 1989 *Phys. Fluids A* **1**, 625.

Newell, A. C. 1974 *Lect. Appl. Math.* **15**, 157.

Preisser, F., Schwabe, D., Scharmann, A. 1983 *J. Fluid Mech.* **126**, 545.

Rupp, R., Muller, G., & Neumann, G. 1989 *J. Crystal Growth* **97**, 34.

Sen, A. K. & Davis, S. H. 1982 *J. Fluid Mech.* **121**, 163.

Shen, Y., Neitzel, G. P., Jankowski, D. F., & Mittelmann, H. D. 1990 *J. Fluid Mech.* **217**, 639.

Smith, M. K. 1986 *Phys. Fluids* **29**, 3182.

Smith, M. K. 1988 *J. Fluid Mech.* **194**, 391.

Smith, M. K. 1990 *J. Fluid Mech.* **217**, 469.

Smith, M. K. & Davis, S. H. 1983a *J. Fluid Mech.* **132**, 119.

Smith, M. K. & Davis, S. H. 1983b *J. Fluid Mech.* **132**, 145.

The George W. Woodruff School of Mechanical Engineering
Georgia Institute of Technology   Atlanta, GA    30332-0405

P H STEEN[+]
# Capillary containment and collapse in low gravity: dynamics of fluid bridges and columns

**1. Introduction.** In the low-gravity environment of a space-laboratory, without the stabilizing influence of the gravity of our experience, free or partially-free bodies of liquid must be stabilized by other means. Surface tension is the obvious candidate. The influence of motion on the containment by surface tension is important to a wide variety of processing applications (materials and/or chemical). Furthermore, motion sometimes leads to surprising configurations.

We study containment and collapse by means of two contrasting questions: can motion enhance the stability of a configuration (containment) and what motions are generated when a configuration is unstable (collapse)? The presentation splits naturally along these lines. Both questions are related to classical problems in applied mathematics but with new twists; the intent is to trace the thread of the story without any attempt at a comprehensive survey.

Stable containment of a fluid body by surface tension (i) in the absence of gravity and other body forces, (ii) in the absence of surface traction (tangential) and (iii) without contact with a solid boundary occurs only for spherical configurations, a classical result. We show that if conditions (ii) and (iii) are relaxed, the capillary instability can be suppressed (Section 2). In particular, the instability of a long cylindrical interface that would otherwise (typically) lead to breakup of the liquid column can be stabilized by certain motions of the liquid underlying the interface.

Collapse occurs whenever stable containment fails. During the collapse process there is a change in topology of the body and predicting the evolution of the interface can be a difficult nonlinear problem. We describe one of the simplest of such problems - for the collapse of a soap-film which bridges two coaxial endrings - and report observations from experiment (Section 3).

## 2. Containment: Shear Stabilization

**Background.** A cylindrical column of motionless fluid influenced solely by surface tension is unstable to infinitesimal disturbances of wavelength $\lambda$ longer than its circumference $2\pi a$ (Figure 1a). This is a classical result [1,2].

The instability is "geometric" in the sense that the critical wavelength is independent of the surface tension. Indeed, the result can be obtained from an energy stability argument which compares surface energies (or, equivalently, surface areas given a uniform interfacial tension) of disturbed and undisturbed static states. By such arguments, it is clear that the same critical wavelength will result if a static annulus of liquid coats a solid rod (Figure 1b). Furthermore, even if the solid boundaries are introduced in a way that constrains the liquid, say with fixed contact lines (circles) as shown in the bridge of Figure 1c, then the same stability limit again occurs [3].

Our discussion so far has been restricted to cylindrical interfaces. The critical wavelength depends on geometry only. In contrast, the rate of growth of disturbances and the wavelength of the fastest growing disturbance (unstable configurations) will depend on the presence and arrangement of the solid boundaries, the interfacial tension, and the nature of the inner and outer fluids through their properties such as viscosities. In order to suppress the geometric instability of the fluid column with fluid motion, all of these factors will come into play.

It turns out that the simplest demonstration of shear-stabilization occurs in the context of an annulus of viscous liquid whose interface is sheared and otherwise bounded by a passive ambient. We formulate the problem in the context of Figure 1b, cite some results particular to this formulation [4] and summarize the results for other related systems.

**Formulation.** Figure 2 shows a sketch of the system geometry and a representative of the class of solutions (base states) whose stability is to be tested. A viscous (incompressible) liquid occupies an axially unbounded region $\Omega$ whose (outer) boundary $\Gamma$ (with outward normal $\mathbf{n}$) forms an interface with an ambient fluid. The liquid flows along a rigid rod (inner boundary $\Gamma_{rod}$) driven by some combination of axial pressure and shear forces. The pressure and shear forces may oppose one another (Figure 2) or may act in concert. The applied shear stress is imposed as a traction vector in a (unit) direction $\mathbf{t}$ tangential to the deformed interface in the positive z-direction, which is well-defined provided deformations are small (we restrict to these below).

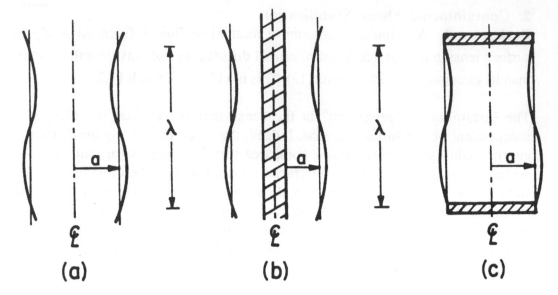

Figure 1

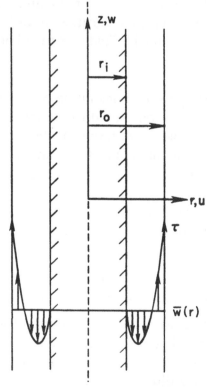

Figure 2

The velocity field $\mathbf{v}$, pressure field p, and the "free" interface

$$\Gamma \equiv (x,y,z \mid 0 < x < \xi \, (y,z,t), \, 0 < y < 2\pi/A, \, -\infty < z < \infty) \qquad (1)$$

are governed by the following initial free-boundary problem for the Navier-Stokes equations (in dimensionless form):

$$\nabla \cdot \mathbf{v} = 0 \qquad\qquad (2a)$$

$$\text{in } \Omega, \quad t > 0$$

$$\partial \mathbf{v}/\partial t + (\mathbf{v} . \nabla)\mathbf{v} = \text{Div}\underline{T} \qquad (2b)$$

$$\mathbf{v} = \mathbf{0} \qquad\qquad \text{on } \Gamma_{\text{rod}}, \, t > 0 \quad (2c)$$

$$(\underline{T}\text{-}\underline{T}_a)\mathbf{n} = 2H\mathbf{n} + RS^{-1}\mathbf{t} \qquad (2d)$$

$$\text{on } \Gamma, \, t > 0$$

$$\mathbf{v} \cdot \mathbf{n} = \mathbf{v} \qquad\qquad (2e)$$

together with an appropriate initial condition for $t = 0$, where

$$\underline{T} \equiv \text{-p} \, \underline{1} + 2S^{-1} \, \underline{D}, \qquad\qquad (3a)$$

$$\underline{T}_a \equiv \text{-p}_a \, \underline{1}, \qquad\qquad (3b)$$

$$\underline{D} \equiv 1/2 \, (\nabla \mathbf{v} + (\nabla \mathbf{v})^t), \qquad\qquad (3c)$$

and H is the mean curvature of the interface and $\mathbf{v}$ is the speed of displacement of $\Gamma$.

These dimensionless equations arise from scaling velocity components with the capillary breakup velocity $W_c$, lengths with d, time $W_c/d$, and pressure and stresses with $\rho W_c^2$. The Table lists the various dimensional and dimensionless parameters.

133

## Table

**dimensional**

| | |
|---|---|
| $(r,\theta,z)$ | cylindrical coordinates |
| $d$ | thickness of liquid annulus (undisturbed) |
| $r_0 - d$ | radius of rigid rod |
| $\sigma$ | surface tension |
| $\mu$ | viscosity |
| $\rho$ | density |
| $\tau$ | applied shear stress |
| $P_a$ | applied ambient pressure |
| $W_c = (\sigma/d\rho)^{1/2}$ | velocity scale (capillary) |
| $W_s = \tau d/\mu$ | velocity scale (shear) |

**dimensionless**

| | |
|---|---|
| $(x,y,z) = (r/d - (1-A)/A,\ \theta/A,\ z/d)$ | coordinates |
| $(u,v,w)$ | velocity components |
| $A = d/r_0 \quad (0 \leq A \leq 1)$ | thickness ratio |
| $R = \rho d W_s/\mu = \rho d^2 \tau/\mu^2$ | shear (Reynolds) number |
| $S^{1/2} = \rho d W_c/\mu = (\rho d\sigma)^{1/2}/\mu$ | capillary (Reynolds) number |
| $T = d\overline{w}/dx\,(0)$ | slope of velocity profile at rod |

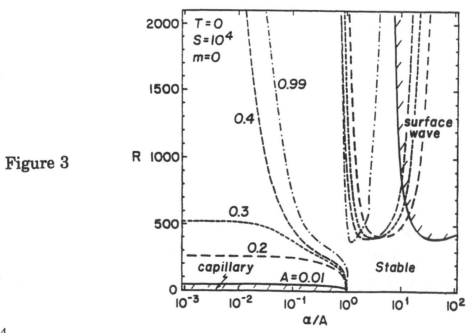

Figure 3

**Basic States.** The governing equations admit a class of steady solutions (denoted with overbars) possessing a cylindrical interface. The flow field (rectilinear) is generated by the constant axial pressure gradient $\overline{p}_z$ and the shear stress $RS^{-1}$ at the interface[1],

$$\overline{\xi}(y,z) = 1, \ 0 < y < 2\pi/A, \ |z| < \infty, \tag{4a}$$

$$\overline{p}(z) - p_a = 1, \tag{4b}$$

$$\overline{u} = \overline{v} = 0 \tag{4c}$$

$$\overline{w}(x) = (\overline{p}_z S^{1/2}/ 4A^2) \{A^2 x^2 + 2Ax (1-A) + 2 (1-2AR/\overline{p}_z S) \ln [(1-A)/(Ax + 1-A)] \} \tag{4d}$$

It is convenient to eliminate the control parameter $\overline{p}_z$ (the curvature of the velocity profile) in terms of the slopes of the velocity profile at the inner boundary,

$$T = d\overline{w}/dx \ (0), \tag{5}$$

through the relationship,

$$\overline{p}_z = 2 S^{-1/2} (2-A)^{-1} [RS^{-1/2} - (1-A) T]. \tag{6}$$

The class of base states is a 3-parameter family of profiles which for fixed geometry (A) depends on the slopes of the velocity profile at the free surface ($RS^{-1/2}$) and at the rigid rod (T).

**Results and Discussion.** The stability of the base state is determined by a standard (temporal) linear analysis where infinitesimal disturbances assume the form,

$$(\hat{v}(x), \hat{p}(x), \hat{\xi}) \exp[i(\alpha z + \beta y - \lambda t)] \ ,$$

------

[1]Note that ambient pressure must have a linear variation to be an exact solution; see [5], [6] for a discussion of the conditions under which this form of the assumed ambient may be a good approximation.

where $\alpha$ and $\beta \equiv mA$ are the axial and azimuthal wavenumbers (real) and $\lambda = \omega + i\,\gamma$. Here, $\omega$ is the radian frequency and $\gamma$ is the growth rate. By restricting $\alpha$ to be real we preclude disturbances which grow in the axial direction consistent with our focus on bridges as distinguished from jets.

Solutions of the above two-point boundary-value problem are summarized by the neutral surfaces (locus of $\gamma = 0$) which separate stable from unstable regions in parameter space. Figure 3 (from [4]) illustrates the main result that disturbances of all wavenumbers can be stable for sufficiently thin films (small A). Indeed, there is a window of moderate applied shear below which the interface is unstable to the classical capillary instability and above which it is susceptible to a 'surface wave' instability. The shear stabilizes the otherwise unstable capillary disturbances.

A check on the stability of the non-axisymmetric modes shows that the axisymmetric mode (m = 0) is the most unstable for thin films [4]; stabilization may be physically realizable. Further exploration shows that when the pressure gradient associated with the base flow is small or absent, stabilization can occur for annular gaps up to nearly a third of the radius of the undisturbed interface. Counterflow at the wall tends to close the window (destabilize) while coflow tends to widen the window (stabilize). The physics involves a stabilizing influence of inertia that for certain layers can swamp the destabilizing effects of the capillary pressure [7,8].

Shear stabilization has been predicted in a variety of related contexts including film flow down the inside of a tube wall [7], thermocapillary driven annular film flow where it is shown that thermal effects do not destroy stabilization [8], as well as the core-annular (Poiseuille flow) of two immiscible liquids along a pipe [9,10].

Thermocapillary stabilization may explain the apparent observation of a very long stable interface (nearly twice the static limit) in a molten indium sample taken from a float zone experiment in a space laboratory (STS Flight 41D, September 1984 [11]). Furthermore, shear stabilization may explain the (limited) success of transporting heavy crude oil through pipelines by using a less viscous liquid to lubricate the pipe wall. A stable interface is the key to reducing the required pumping power [12].

We return to Figure 1 to summarize. Shear applied to the liquid column (Figure 1a) damps the axisymmetric instability but not to the extent that waves of all lengths decay [4]. On the other hand, the liquid column with central rod (Figure 1b) can be completely stabilized as just discussed. As for the system with finite endwalls (Figure 1c), we are currently examining its prospects of stabilization through both stability calculations and experiment.

136

## 3. Collapse: Soap-film Bridge Prototype.

**Background.** The soap-film bridge is shown in schematic in Figure 4. The film stretches between two open endrings where the contact lines are held fixed. For sufficiently thin films (on earth), the force of capillarity dominates gravity. In contrast to the closed endwall system (Figure 1c) discussed above, the open endrings allow pressure to equilibrate across the interface.

As far as equilibrium states are concerned, the soap-film may be idealized as a mathematical surface. The stability of stationary states can be obtained through minimization of the free-energy functional based on surface energy. The results belong to the classical literature of the calculations of variations [cf. 13]. The Young-LaPlace equation predicts minimal surfaces as equilibrium shapes and the connected minimal surfaces are surfaces of revolution generated by the catenary (catenoids). The disconnected minimal surfaces are circular discs which cover the two endrings. For $L/D < (L/D)_c$ ( $= 1.3255$) there are two catenoid solutions, one stable (local minimum) and the other unstable, which coalesce at $L/D = (L/D)_c$. The disconnected solution exists for all values of $L/D$. There is an aspect ratio where the energy of the disconnected solution just equals that of the stable catenoid; for larger separations the disconnected solution has lower energy. Figure 5 summarizes the set of solutions as measured by the energy (normalized to the covered endring solution) and parameterized by the aspect ratio. The unstable catenoid is the broken line. The circles represent experimentally measured values.

In summary, just as for the liquid column, the stationary soap-film can exist as a stable connected state only for separation of the endrings up to a critical value. In contrast, however, the instability of the soap-film occurs as a "saddle-node" bifurcation whereas the instability of the column, occurs as a "pitchfork" bifurcation. In particular, there exists no local branch of equilibrium states beyond the critical length for the soap-film collapse.

**Observations.** Connected equilibrium configurations have a minimum radius d (neck). By dimensionless analysis the neck radius can only depend on the aspect ratio

$$d/D = f(L/D).$$

Since the collapse process is driven by the free-energy difference between initial and final states, and since the soap-film displaces a certain volume of air during the process, it is reasonable to form a time-scale based on surface-tension $\sigma$ and the density of air $\rho$. The simplest functional description might be written,

$$d/D = f(\sigma t^2/\rho D^3, (L/D)_c).$$

137

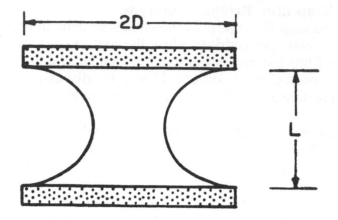

Figure 4

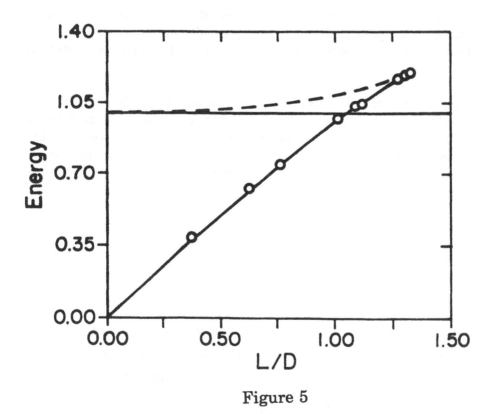

Figure 5

Furthermore, one might expect a similarity solution of the inviscid hydrodynamic equations to be appropriate in limiting cases [e.g., 14]. Unfortunately, the behavior of the system seems not to follow such a simple scenario. In particular, several regimes with different exponents of time are evident during collapse.

Experiment confirms the critical value of the aspect ratio to good accuracy. Furthermore, the dynamic sequence of states traversed during collapse (less than $10^{-1}$ seconds) is very reproducible. Four regimes or phases of rupture are identified. During the first two phases (0.04 second) the shape remains axisymmetric and the diameter of neck decreases monotonically from the diameter of the catenoid neck. Three of the four phases are indicated on Figure 6 which plots the displacement of the interface during a fixed time interval (1/60 second) versus the position of the interface as measured from video recordings of many collapse experiments. Phase one and two are distinguished by a log-log plot (not shown) while phase three is clearly evident from Figure 6.

The sequence of shapes exhibited during the first phase can be adequately predicted by a quasistatic theory of constrained equilibria which, however, cannot predict the exponent of the time-dependence [15].

The second phase is marked by an inward acceleration of the film and shows a steeper exponent for the time-dependence. Both the first and second phases seem independent of the nature of the ambient disturbances which spontaneously destabilize the bridge although they may turn out to be quite sensitive to details of end conditions.

The third phase is characterized by the formation of a small bubble on the axis midway between the endrings. The beginning of this phase is marked by the change from a single minimum in radius at the midplane to a pair of minima which approach the axis as the bubble seals off. To either side of the bubble the film has collapsed completely to form a thin cylindrical thread of liquid bridging the two film caps, each of which recedes toward the respective endrings. The liquid in the cylindrical thread forms an axial jet of sufficient strength to locally distort the shape of the receding caps. The thread subsequently breaks up into droplets, succumbing to a capillary jet instability. This marks the end of the third phase (0.001 second).

During the fourth phase, the caps (originally hemispherical-like) relax to the flat planar discs which cover the two endrings. Capillary waves are clearly evident as the hemispheres flatten to the planar state.

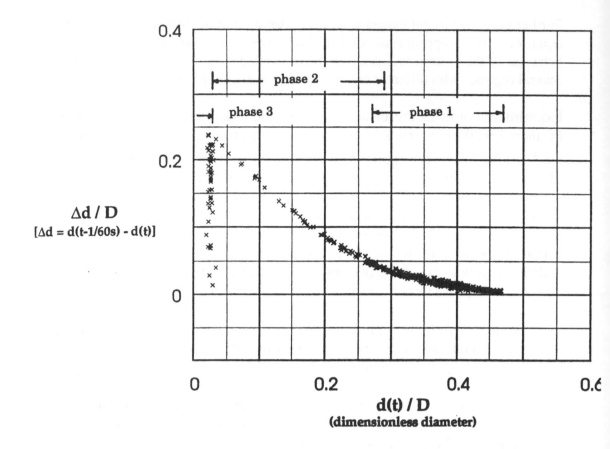

Figure 6

**Discussion.** Previous studies of rupture can be distinguished from the soap-film bridge breakup we observe [cf. 16, 17]. For those, typically, a large disturbance (e.g., a spark) pierces a planar film and the growth in time of the hole is observed. In contrast, we observe the collapse of a hole. Taylor and Michael [18] view the catenoid as a model for a hole in a sheet of liquid using the critical aspect ratio as a criteria for hole formation. For large enough holes, $L/D < (L/D)_c$ where here L represents the thickness of the sheet and D a measure of the hole diameter, the hole persists while for small enough holes, $L/D > (L/D)_c$, the holes have insufficient surface energy to survive.

In hole formation, large surface energies can be liberated while in hole collapse only 20% of the system's energy is used up. The physics of collapse is expected to be more complicated if only because the system begins and ends at equilibrium. On the other hand, both processes have features in common. The acceleration observed in phase two extrapolates to infinite velocities at $d = 0$, and may be related to the singularities described in [19]. Of course, phase three and bubble formation represent a regime of different physics which takes over when velocities are sufficiently large.

The breaking of axisymmetric liquid bridges in a viscous bath (neutrally buoyant) shares features with the collapse of the soap-film bridge. In particular, inviscid analysis has been successful in accounting for dynamics for times short of the collapse point [20], [21]. Furthermore, the bubbles observed in phase three seem to have counterparts in the breakup of liquid bridges [22].

In summary, the collapse of a soap-film bridge shows a rich variety of behaviors, perhaps unexpected of a free-boundary problem driven by surface tension.

**Acknowledgements.** The photographs (presentation) and figures pertaining to bridge collapse are from the work of Dr. Steven A. Cryer [15], with the exception of Figure 6, which has been contributed by Mr. Brian Lowry. The shear-stabilization results are from the work of Dr. Matthew Russo [7] with contributions from Dr. Hendrick Dijkstra. Primary support of this work is through NASA Grant No. NAG3-801.

The Alexander von Humboldt Foundation and the Institut für Reaktorbauelemente, Kernforschungszentrum, Karlsruhe, West Germany are gratefully acknowledged for their partial support of the author's travel to this meeting.

# References

1. J.A.F. PLateau, Experimental and Theoretical Researches on the Figures of Equilibrium of a Liquid Mass..., (translation) in <u>Annual Reports of the Smithsonian Institution</u> (1863-1866).

2. Lord J.W.S. Rayleigh, On the Instability of Jets, <u>Proc. London Math. Soc.</u> <u>10</u>, 4 (1879).

3. R.D. Gillette and D.C. Dyson, Stability of Fluid Interfaces of Revolution Between Equal Solid Circular Plates, <u>Chem. Eng. J.</u> <u>2</u> 44 (1971).

4. M.J. Russo and P.H. Steen, Shear Stabilization of the Capillary Breakup of a Cylindrical Interface, <u>Phys. Fluids A1</u> (12), (1989).

5. M.K. Smith and S.H. Davis, The instability of sheared liquid layers, <u>J. Fluid Mech. 121</u>, 187 (1982).

6. J.W. Miles, The hydrodynamic stability of a thin film of liquid in uniform shearing motion, <u>J. Fluid Mech 8</u>, 593 (1960).

7. M.J. Russo, PhD. Dissertation, Cornell University, 1990.

8. H.A. Dijkstra and Paul H. Steen, Thermocapillary stabilization of the capillary break-up of an annular film of liquid, <u>J. Fluid Mech. 229</u>, 205 (1991).

9. L. Preziosi, K. Chen, and D.D. Joseph, <u>J. Fluid Mech. 201</u>, 323 (1989).

10. H. Hu and D.D. Joseph, Lubricated pipelining: stability of core annular flow: part 2, <u>J. Fluid Mech. 205</u>, 359 (1989).

11. S.P. Murphy, J.J. Hendrick, M.J. Martin, R.W. Grant, and M.D. Lind, Floating-zone Processing of Indium in Earth-Orbit, <u>Mat. Res. Soc. Symp. Proc., 87</u> (1987).

12. T.W.F. Russel, and M.E. Charles, The effect of the less viscous liquid in the laminar flow of two immiscible liquids, <u>Cand. J. Chem. Eng., 39</u> (1959).

13. C.Caratheodory, <u>Calculus of Variations</u>, (First English Edition) Holden-Day, Inc. 1965.

14. J.B. Keller and M.J. Miksis, Surface Tension Flows, <u>SIAM J. Appl. Math. 43</u> (2) (1983), 268-277.

15. S.A. Cryer, PhD. Dissertation, Cornell University, 1990.

16. W.E. Ranz, Some experiments on the dynamics of liquid films, <u>J. Appl. Phys. 30</u>, 1950 (1959).

17. W.R. McEntee and K.J. Mysels, The Bursting of Soap Films. I. An Experimental Study, <u>J. Phys. Chem. 73</u>, 33018-3028 (1969).

18. G.I. Taylor and D.H. Michael, On Making Holes In a Sheet of Fluid, <u>J. Fluid Mech.</u>, <u>58</u>, 625-667 (1973).

19. J.B. Keller, Breaking of Liquid Films and Threads, <u>Phys. Fluids 26</u> (12), 3451-3453 (1983).

20. J. Meseguer, The Breaking of Axisymmetric Slender Liquid Bridges, <u>J. Fluid Mech. 130</u>, 123-151 (1983).

21. D. Rivas and J. Meseguer, One-Dimensional Self-Similar Solution of the Dynamics of Axisymmetric Slender Liquid Bridges, <u>J. Fluid Mech. 138</u>, 417-429 (1984).

22. J. Meseguer and A. Sanz, Numerical and Experimental of the Dynamics of Assymmetric Slender Liquid Bridges, <u>J. Fluid Mech. 153</u>, 83-101 (1985)

Paul H. Steen
School of Chemical Engineering, Cornell University, Ithaca, New York 14853

# Chemical and biological reactions I

M BERTSCH, D HILHORST AND J HULSHOF

# On a free boundary problem arising in detonation theory: convergence to travelling waves

## 1. INTRODUCTION.

In this note we consider the free boundary problem

$$(P_1) \begin{cases} u_t = u_{xx} + uu_x & t > 0 \,, x \neq \varsigma(t) \,; \\ u(\varsigma(t)^-, t) = u(\varsigma(t)^+, t) = q & t > 0 \,; \\ u_x(\varsigma(t)^-, t) - u_x(\varsigma(t)^+, t) = 1 & t > 0 \,; \\ u(x, 0) = u_0(x) & x \in \mathbf{R} \,; \\ \varsigma(0) = \varsigma_0 \,, \end{cases}$$

where $q$ is a positive constant, $\varsigma_0$ is a given real number, $u_0$ a given initial function satisfying $u_0(\varsigma_0) = q$, and $\varsigma(t)$ and $u(x,t)$ are the unknown functions to be found. This problem arises in combustion theory and was introduced by Stewart in [12] and by Ludford and Oyediran in [10]. It is a simple model of detonation waves where the reaction is supposed to occur at the detonation front $x = \varsigma(t)$, and where $q$ is the ignition temperature.

We consider three configurations for the initial function $u_0$.
(i) $0 \leq u_o \leq A$ for some constant $A > q$, $u_0(x) < q$ for $x < \varsigma_0$ and $u_0(x) > q$ for $x > \varsigma_0$.
(ii) $0 \leq u_o \leq q$ and $\varsigma_0 = \sup\{x < \varsigma_0 : u_0(x) < q\}$.
(iii) $0 \leq u_o \leq q$ and $u_0(x) < q$ for all $x \neq \varsigma_0$.

In Section 2 we consider the first case and show, under some additional assumptions on the initial data, that Problem $(P_1)$ has a unique solution $(u, \varsigma)$ which is such that $0 \leq u \leq A$, $u(x,t) < q$ for all $x < \varsigma(t)$ and $u(x,t) > q$ for all $x > \varsigma(t)$.

The second case, in which Problem $(P_1)$ reduces to the "one-phase" problem

$$(P_2) \begin{cases} u_t = u_{xx} + uu_x & t > 0 \,,\ x < \varsigma(t) \,; \\ u(\varsigma(t),t) = q & t > 0 \,; \\ u_x(\varsigma(t),t) = 1 & t > 0 \,; \\ u(x,0) = u_0(x) & x < \varsigma_0 \,; \\ \varsigma(0) = \varsigma_0 \,, \end{cases}$$

is considered in Section 3. Under some mild regularity asssumptions on $u_0$, we show the existence and uniqueness of the solution of Problem $(P_2)$.

In the third case, we doubt whether Problem $(P_1)$ is well-posed: if we omit the convection term $uu_x$ in the parabolic equation for $u$, and take an initial profile $u_0$ which is symmetric around $\varsigma_0$, then the well-posedness would imply that $\varsigma(t) \equiv \varsigma_0$ and that the solution remains symmetric around $\varsigma_0$. But then the jump condition in $u_x$ across the free boundary reduces to $u_x(\varsigma_0,t) = 1/2$ which makes this problem overdetermined.

One could think of solving Problem $(P_1)$ by means of a fixed point method based on an iterative procedure of the form: given $\varsigma$ solve for $u$, then compute $\varsigma$ from $u$ and so on. Unfortunately, $\varsigma$ appears only implicitly in the free boundary conditions and it is very hard to carry out the second step. The only explicit formula one has is

$$\varsigma'(t) = - \lim_{x \uparrow \varsigma(t)} \left\{ \frac{u_{xx}(x,t)}{u_x(x,t)} + u(x,t) \right\} \quad,$$

which is difficult to work with because of the second order term $u_{xx}$. This motivated us to look for weak formulations of the problems $(P_1)$ and $(P_2)$ in which the free boundary does not appear. The free boundary is characterised a posteriori as the level set of $q$ of the weak solution $u$. We were able to do this for the first two cases but not for the third.

In Section 4 we study the large time behaviour of these solutions $(u,\varsigma)$. It turns out that whenever the problem has a unique (up to translation) travelling wave solution, any solution converges to a travelling wave.

In [4] Brauner, Noor Ebad and Schmidt-Lainé have studied the local exponential stability of travelling wave solutions of Problem $(P_2)$ without the convection term. For the results on the two-phase problem we refer to [1] and [2], and the one-phase problem is dealt with in [7]. Finally we mention that in [11] a related Stefan problem with super-cooling is investigated. Integrating this problem with respect to the space variable one obtains our one-phase problem without the convection term.

## 2. WELL-POSEDNESS OF THE TWO-PHASE PROBLEM.

In this section we assume that $u_0$ satisfies the following hypothesis:

H1. $u_0 \in C^{0,1}(\mathbf{R})$, $0 \le u_0 \le A$, $u_0' + H(u_0 - q) \in C^{0,1}(\mathbf{R})$, where $H$ is the Heaviside function, $u_0 - AH \in L^1(\mathbf{R})$, and, for some $\delta > 0$ and $\varepsilon > 0$, $u_0' \ge \delta$ in the set $(a,b) = \{x \ne \varsigma_0, |u_0(x) - q| < \varepsilon\}$, $u_0(x) < q - \varepsilon$ for $x < a$ and $u_0(x) > q + \varepsilon$ for $x > b$;

and associate to Problem $(P_1)$ the problem

$$(Q_1) \begin{cases} u_t = u_{xx} + uu_x + \{H(u-q)\}_x & t > 0 \, , \, x \in \mathbf{R} \, ; \\ u(x,0) = u_0(x) & x \in \mathbf{R} \, , \end{cases}$$

In [1] we prove the following results:

There exists a unique (weak) solution u of Problem $(Q_1)$. Moreover there exists a function

$$\varsigma \in C^{0,1}([0,\infty)) \cap C^1((0,\infty)) \quad ,$$

such that

$$\{(x,t) \in \mathbf{R} \times \mathbf{R}^+ : u(x,t) = q\} = \{(x,t) : t > 0, x = \varsigma(t)\} \quad .$$

The pair $(u,\varsigma)$ is a solution of Problem $(P_1)$.

The uniqueness proof for Problem $(Q_1)$ is rather standard and so is the idea of the existence proof: we regularize the Heaviside function in the parabolic equation. The proof of the convergence of the regularized solution to a weak solution is not standard. We really need precise information about the level curves, hence the strong assumptions on the initial data. Formally these level curves are defined by

$$u(X(c,t),t) = c \quad ,$$

for c near q where $X_t$ turns out to satisy the equation

$$(X_t)_t = \left\{ \frac{(X_t)_c}{X_c^2} \right\}_c \quad .$$

Using the assumptions we show that for the regularized solution $u_x$ is positive near the level set of $q$. A maximum principle argument permits us to conclude that $X_t$ is bounded near that same level set. One can then show that the level set of $q$ is really a level curve $x = \varsigma(t)$. Near this curve $u_x$ can be bounded away from zero, from which we finally deduce that $\varsigma(t)$ is continuously differentiable for $t > 0$.

## 3. WELL-POSEDNESS OF THE ONE-PHASE PROBLEM.

In this section we assume that $u_0$ satisfies the following hypothesis:

H2. $u_0 \in C((-\infty, \varsigma_0]) \cap L^1(-\infty, \varsigma_0)$, $0 \le u_0 \le q$, $u_0(x) \to 0$ as $x \to -\infty$, $u_0(\varsigma_0) = q$, $\varsigma_0 = \sup\{x < \varsigma_0 : u_0(x) < q\}$;

and associate to Problem $(P_2)$ the elliptic-parabolic problem

$$(Q_2) \begin{cases} c(u)_t = u_{xx} + c(u)c(u)_x & t > 0 \, , \, x < R \, ; \\ u_x + \frac{1}{2}c(u)^2 = 1 + \frac{1}{2}q^2 & t > 0 \, , \, x = R \, ; \\ c(u(x,0)) = v_0(x) & x < R \, , \end{cases}$$

where

$$c(s) = \min(s,q) \quad ,$$

148

$v_0 \equiv u_0$ on $(-\infty, \varsigma_0]$, and $v_0 \equiv q$ on $(\varsigma_0, R)$. Note that for $u < q$ the differential equation is the same parabolic equation as before, while for $u > q$ it reduces to the elliptic equation $u_{xx} = 0$. Related problems on bounded domains have been studied by several authors, see e.g. [6,5,9,8,3]. Formally a solution of Problem $(P_2)$ can be viewed as a (weak) solution of $(Q_2)$ provided that $R > \varsigma$ if it is extended by setting

$$u(x,t) = q + x - \varsigma(t) \quad ,$$

for $x > \varsigma(t)$.

Existence and uniqueness of a weak solution $u$ of $(Q_2)$ are established in [7] following [6,9]. However this weak solution may be strictly less than $q$ on the whole of $(-\infty, R]$ for some $t > 0$. In order to avoid this possibility we assume that

$$R \geq \varsigma_0 + q$$

Under this condition we show in [7] along the lines of [8] that there exists a continuous interface $x = \varsigma(t)$ separating the regions $\{u < q\}$ and $\{u > q\}$, and that $(u, \varsigma)$ is a solution of $(P_2)$. It can then be shown that $u_x$ is strictly positive in a neighbourhood of the interface and consequently the following transformation makes sense:

$$p = u_x \quad , \quad \xi = u \quad , \quad \tau = t \quad ,$$

For $p = p(\xi, \tau)$ we obtain

$$p_\tau = p^2 (p_{\xi\xi} + 1) \quad ,$$

and the Neumann condition on the free boundary transforms into

$$p(q, \tau) = 1 \quad .$$

Standard regularity theory allows us to conclude that $p$ is smooth up to $\xi = q$. Since

$$\frac{u_t}{u_x} = p_\xi + \xi \quad ,$$

this implies that the level curves near the free boundary and in particular the free boundary itself are smooth for positive $t$. Hence $u$ is also smooth up to the free boundary.


## 4. CONVERGENCE TO TRAVELLING WAVES.

First we consider the two-phase problem (case (i) in Section 1). There exists a unique travelling wave velocity

$$\omega = \frac{1}{A} + \frac{A}{2} \quad ,$$

and a corresponding travelling wave solution of the form

$$u(x,t) = U(x + \omega t) \quad ,$$

149

unique up to translation, if and only if $q$ and $A$ satisfy the travelling wave condition

$$\frac{2}{A} < q < A \quad .$$

If this condition is satisfied we prove in [2] that the solution $u$ converges to a translate of the travelling wave, and that the difference of the corresponding interfaces tends to zero as $t \to \infty$. In future work we shall come back to the case where the travelling wave condition is not satisfied.

For the one-phase problem (case (ii) in Section 1) there always exists a unique travelling wave velocity

$$\omega = \frac{1}{q} + \frac{q}{2} \quad ,$$

and a unique travelling wave profile $U$. The convergence of every solution to a travelling wave as $t \to \infty$, as well as the convergence of the difference of the interfaces to zero are established in [7].

In both cases the translate is determined by the condition that at $t = 0$ the integral of the difference of $u_0$ and the travelling wave equals zero. The proofs are based on the $L^1$-contractivity of the solution operators corresponding to the problems $(P_1)$ and $(P_2)$, and on comparison principle arguments.

Let us now come back to the the two-phase problem in the case that

$$0 < q < \frac{2}{A} \quad .$$

Although a "global" travelling wave does not exist, the asymptotic behaviour of the solution as $t \to \infty$ can be described by means of two travelling waves: the travelling wave $U^-$ of the one-phase problem described just above with velocity

$$\omega^- = \frac{1}{q} + \frac{q}{2} \quad ,$$

and the travelling wave $U^+$ of Burger's equation with $U^+(-\infty) = q$ and $U^+(+\infty) = A$, which has velocity

$$\omega^+ = \frac{q + A}{2} < \omega^- \quad .$$

Then in the travelling wave coordinates corresponding to $\omega^-$, $\min(u, q)$ converges to a translate of $U^-$, and in the travelling wave coordinates corresponding to $\omega^+$, $\max(u, q)$ to a translate of $U^+$.

## REFERENCES

[1] BERTSCH, M., D. HILHORST and Cl. SCHMIDT-LAINE, *The well-posedness of a free boundary problem arising in combustion theory*, Preprint Ecole Norm. Sup. Lyon no. 21 (1989).

[2] BERTSCH, M. and D. HILHORST, *On a free boundary problem arising in combustion theory : The large time behaviour*, in preparation.

[3] BERTSCH, M. and J.HULSHOF, *Regularity results for an elliptic- parabolic free boundary problem*, Trans. Amer. Math. Soc. 297 (1986) 337-350.

[4] BRAUNER, C.M., S. NOOR EBAD and Cl. SCHMIDT-LAINE, *Sur la stabilité d'ondes singulières en combustion*, C.R. Acad. Sci. Paris 308 (1989) 159-162.

[5] VAN DUYN, C.J., *Nonstationary filtration in partially saturated media: Continuity of the free boundary*, Arch. Rat. Mech. Anal. 79 (1982) 261-265.

[6] VAN DUYN, C.J. and L.A. PELETIER, *Nonstationary filtration in partially saturated porous media*, Arch. Rat. Mech. Anal. 78 (1982) 173-198.

[7] HILHORST, D. and J. HULSHOF, *An elliptic-parabolic problem in combustion theory: Convergence to travelling waves*, Preprint Univ. Paris-Sud 90-03 (1990).

[8] HULSHOF, J., *An elliptic-parabolic free boundary problem : Continuity of the interface*, Proc. Roy. Soc. Edinburgh, 106A (1987) 327-339.

[9] HULSHOF, J. and L.A. PELETIER, *An elliptic-parabolic free boundary problem*, J. Nonlinear Analysis TMA, 10 (1986) 1327-1346.

[10] LUDFORD, G.S.S. and A.A. OYEDIRAN, *Numerical aberrations in a Stefan problem from detonation theory*," Numerical Simulation of Combustion Phenomena", R. Glowinski, B. Larrouturou and R. Temam Eds, Lecture Notes in Physics 241, Springer 1985.

[11] RICCI, R. and XIE WEIQING, *On the stability of some solutions of the Stefan problem*, to appear in Eur. J. Appl. Math.

[12] STEWART, D.S., *Transition to detonation in a model problem*, J. Méc. Théor. Appl. 4 (1985) 103-137.

by

**M.Bertsch**

Dipartimento di Matematica
Università di Torino
Via Principe Amedeo 8
10123 Torino, Italy


**D.Hilhorst**

Laboratoire d'Analyse Numérique
CNRS et Université Paris-Sud
91405 Orsay, France


**J.Hulshof**

Mathematisch Instituut
Rijksuniversiteit Leiden
P.O.Box 9512     2300 RA Leiden, The Netherlands

H BYRNE AND J NORBURY
# Catalytic converters and porous medium combustion

**Physical Interpretation.** We study an implicit free boundary value problem (BVP)(1)-(4) which derives from a model of porous medium combustion [4] when a certain asymptotic limit is taken. In this limit there is negligible solid consumption and the solid and gas phases are in thermal equilibrium. Our model may describe the action of a catalytic convertor [6], a device used in motor-vehicles to reduce pollution effects. As the engine fumes pass through a porous catalyst a combustion reaction converts them into harmless waste products. The dependent variables $u,g$ are interpreted as temperature and gaseous concentration respectively, whilst $r$ denotes the reaction rate. The step-function factors in $r$, see (3), signify that the reaction can only be sustained when both the temperature and gas concentration exceed critical values. The key parameters are $\lambda$ and $\mu$ where

$$\lambda = \frac{L^2}{u_c} \text{ and } \mu = \frac{La}{vg_a} \, ,$$

$L$ is the length of the catalyst; $u_c$ is the critical switching temperature, below which the chemical reaction cannot be sustained; $a$ is the ratio of gaseous to solid fuel consumption; $v$ is the inlet gas velocity; $g_a$ is the inlet gas concentration.

The analysis concentrates on the existence of steady-state solutions of (1)-(4), and examines their stability with respect to small time-dependent perturbations. Of particular interest are the length $s(t)$ of the free boundary, the maximum temperature $u(0,t)$, and their dependence on the parameters $(\lambda,\mu)$. We show that $\lambda > \mu$ is a necessary condition for the existence of steady-state solutions. In the original variables this yields $L > au_c/vg_a$ — we have a lower bound on the length of a convertor below which steady solutions cannot be sustained. Using a combination of numerical and analytical techniques we conclude by showing that for two asymptotic limits both stable and unstable solutions to the problem exist. The stable solutions correspond to reactions terminated because the gas concentration has been exhausted (G-solutions), whilst the unstable ones represent reactions terminated by the temperature falling below the critical value (U-solutions).

**The Model.** The simplified model we study is the mixed hyperbolic-parabolic system given by

$$\frac{\partial u}{\partial t} = \frac{\partial^2 u}{\partial x^2} + \lambda r \ , \tag{1}$$

$$\frac{\partial g}{\partial x} = -\mu r \ , \tag{2}$$

$$r = H(u-1)H(g)f(u,g) \ . \tag{3}$$

Here $H(.)$ is the Heaviside step-function and $\lambda$ and $\mu$ are positive parameters. The following initial and boundary conditions are imposed on the system:

$$u_x(0,t) = 0 = u(1,t), \ g(0,t) = 1, \ u(x,0) = u_0(x) \text{ a prescribed function.} \qquad (4)$$

We seek non-trivial solutions of the free BVP (1)-(4) for which $\exists \ s(t) \in (0,1)$ such that $r > 0 \ \forall \ x \in (0,s)$ and $r = 0 \ \forall \ x \in (s,1)$. The variables $u$, $u_x$ and $g$ are assumed continuous across $x = s(t)$. The form of $r$ enables us to deduce that at $x = s(t)$ either $(u = 1, g > 0)$ or $(u > 1, g = 0)$. We differentiate between these cases by making the following definitions:

**Definitions:** Given $(\lambda,\mu) \in \Re_+^2$, if there exist functions $(u,g)$ which satisfy equations (1)-(4) we say that $(\lambda,\mu,u,g)$ constitute

- a U-solution if, in addition, $\{u(s(t),t) = 1, \ g(s(t),t) > 0\}$;

- a G-solution if, in addition, $\{u(s(t),t) > 1, \ g(s(t),t) = 0\}$;

- a degenerate U/G-solution if, in addition, $\{u(s(t),t) = 1, \ g(s(t),t) = 0\}$.

**Steady-State Solutions.** Setting $\partial/\partial t = 0$ in equations (1)-(4) leads to the steady-state problem

$$\left. \begin{aligned}
u_x &= -\frac{\lambda}{\mu}(1-g) \\
g_x &= -\mu f(u,g) \\
\Rightarrow \frac{dg}{du} &= \frac{\mu^2}{\lambda} \frac{f(u,g)}{1-g}
\end{aligned} \right\} \quad 0 < x < s \ ,$$

with

$$u_x(0) = 0, \quad g(0) = 1, \quad u_x(s) = -\frac{u(s)}{1-s} = -\frac{\lambda}{\mu}(1-g(s)),$$

and one of the matching conditions stated above *(u(s),g(s) are constants to be determined as part of the solution)*. The above relationship between $u(s)$ and $g(s)$, obtained from continuity of $u_x$ across $x = s(t)$ and solving explicitly for $u$ and $g$ in $s < x < 1$, enables us to deduce the following:

**Lemma 1** *A necessary condition for the existence of steady solutions is* $\lambda > \mu$.

By restricting attention to the class of functions $f(u,g) = u^m$, $m \geq 1$, we can exploit the form of the problem above to obtain a series of algebraic equations which determine the unknowns $u(0)$, $u(s)$, $g(s)$ and $s$. Defining

$$M = u(0), \quad R = \frac{u(s)}{u(0)}, \quad G = 1 - g(s),$$

these identities become

$$G^2 = \frac{2\mu^2}{\lambda(m+1)} M^{m+1}\left(1 - R^{m+1}\right),$$

$$s = \left[\frac{M^{1-m}(m+1)}{2\lambda}\right]^{1/2} \int_R^1 dv\left(1 - v^{m+1}\right)^{-1/2},$$

$$\frac{MR}{G} = \frac{\lambda}{\mu}(1-s),$$

with either $\{R = M^{-1}, 0 < G < 1\}$, for a U-solution, or $\{R > M^{-1}, G = 1\}$, for a G-solution. Asymptotic analysis of these relations is possible for the limiting cases $(0 < s << 1)$ and $(0 < 1 - s << 1)$. For example, we can deduce the existence of U-solutions having

$$0 < s << 1, \ \lambda \sim \frac{1}{s}, \ 0 < \mu < \frac{1}{s}, \ u(0) \sim 1 + \frac{s}{2} \ .$$

Moreover, it can be shown that all valid solutions having $0 < s << 1$ satisfy

$$\lambda \geq \frac{2 + (1-m)s}{2s} \sim \frac{1}{s} \ , \tag{5}$$

provided $0 < |s(1-m)| << 1$. The above analysis was used to construct figure 1 which describes the existence regimes for steady U/G-solutions in $(\lambda,\mu)$ parameter space and leads us to speculate that degeneracy separates (stable) G-solutions from (unstable) U-solutions.

**Fixed-domain Problem.** Using a combination of numerical and stability techniques, we now attempt to verify the claims made above. The need for both numerical accuracy and to simplify the stability analysis as given in [5] forces us to first map the original BVP onto a fixed domain [1]. Introducing new independent variables $(X,T)$, with $T = t$, equations (1)-(4) become

$$\left. \begin{aligned} u_T &= \frac{1}{s^2} u_{XX} + \frac{\dot{s}}{s} X u_X + \lambda f(u,g) \\ g_X &= -s\mu f(u,g) \end{aligned} \right\} \ 0 < X = \frac{x}{s(t)} < 1, \tag{6}$$

$$\left. \begin{aligned} u_T &= \frac{1}{(1-s)^2} u_{XX} + \frac{\dot{s}(2-X)}{(1-s)} u_X \\ g_X &= 0 \end{aligned} \right\} \ 1 < X = \frac{x+1-2s}{1-s} < 2, \tag{7}$$

156

$$u_X(0,T) = 0 = u(2,T), \quad g(0,T) = 1 ,$$

$$[u(X,T)]_-^+ = 0 = [g(X,T)]_-^+ ,$$

$$\frac{u_X^-(1,T)}{1-s} = \frac{u_X^+(t,T)}{s} ,$$

with either $u(1,T) = 1$ or $g(1,T) = 0$ (the superscripts '+' and '–' denote the solution to the right and left of $X = 1$).

The remaining analysis concentrates on the specific case $f(u,g) = u$, chosen for its simplicity. However, guided by the asymptotic analysis above (see (5)), we expect qualitatively similar stability results to occur for other choices of $m > 1$.

**Numerical Results.** The Keller Box Scheme [3] was used to discretise the transformed equations. At each time-step the resulting system of nonlinear equations was solved using Newton iteration. In this way, given initial data $(u(x,0), g(x,0), s(0))$, we trace the evolution of the free boundary for U- and G-solutions. Figure 2 describes results obtained for the case $(\lambda = 25, \mu = 12)$, chosen for comparison with our asymptotic analysis — $\lambda = 25$ implies that the steady-state free boundary lies at $s(0) \sim 0.04 \sim 1/\lambda$. The four curves, $G1(t)$, $U1(t)$, $G2(t)$ and $U2(t)$, in figure 2 trace the evolution of free boundaries whose initial profiles satisfy $0 < S_{G1}(0) = S_{U1}(0) < s(0) < s_{G2}(0) = s_{U2}(0)$. The divergence of $U1(t)$ and $U2(t)$ together with the (minimal) convergence of $G1(t)$ and $G2(t)$ supports the claims made earlier that G-solutions are stable and U-solutions unstable.

**Stability Analysis.** Following [2], we seek time-dependent perturbation from steady-state solutions of the transformed problem of the form

$$u(X,T) = u(X) + \delta e^{\rho T} \phi(X). \tag{8}$$

These perturbations are assumed to be small relative to the steady-state solutions and hence we linearise the perturbed equations, obtaining eigenvalue problems for U- and G-solutions. In each case the problem reduces to the solution of a dispersion relation for $\rho$ in terms of the parameter(s) $\lambda$ (and $\mu$, when $m > 1$). From the form of the perturbations we infer that if $\Re(\rho) > 0$ for any part of the spectrum then the corresponding steady-state is unstable, whilst $\Re(\rho) < 0$ for all modes suggests stability. In this way we examine the stability of U- and G-solutions for the asymptotic limits $\underline{A}$: $(0 < s \ll 1, \lambda \sim 1/s)$ AND $\underline{B}$: $(0 < 1 - s \ll 1, \lambda \sim \pi^2/4)$ mentioned earlier. The results are summarised in the following:

**Theorem 1** *(i) Steady-state G-solutions of (1)-(4) for the asymptotic cases $\underline{A}$ and $\underline{B}$ are stable with respect to small time-dependent perturbations of the form (8). (ii) Steady-state U-solutions of (1)-(4) for the same limits are unstable.*

**Note.** The first author wishes to thank the Science and Engineering Research Council for funding this work, an expanded version of which is to be submitted to the SIAM Journal of Applied Mathematics in 1990.

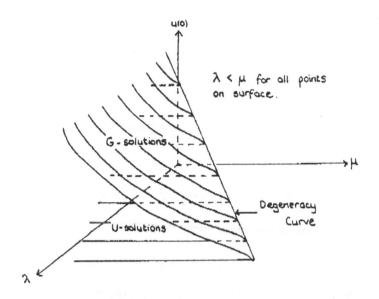

Figure 1: Steady-State Solution Surface; $f(u,g) = u$. The upper, stable G-solution surface meets the lower, unstable U-solution surface on the degeneracy curve.

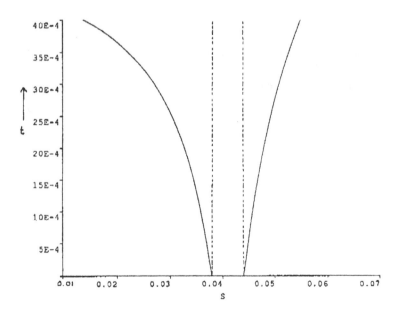

Figure 2: Evolution of Free Boundary; $\lambda = 25$, $\mu = 12$.

## References

[1]    Crank, J., *Free and Moving Boundary Problems*, Oxford, Clarendon Press, 1984.

[2]    Drazin, P.G. and Reid, W.H., *Hydrodynamic Stability*, Cambridge University Press, 1981.

[3]    Keller, H.B., in *Numerical Solution of Partial Differential Equations II* (ed: Hubbard), Academic Press, New York, 1971.

[4]    Norbury, J. and Stuart, A.M., *A Model for Porous-Medium Combustion*, Quart. J. Mech. and Appl. Math. 42, 1989, pp. 159-178.

[5]    Norbury, J. and Stuart, A.M., *Travelling Combustion Waves in a Porous Medium. Part I - Existence, Part II - Stability*, SIAM J. Appl. Math., 48, 1988, pp. 155-169 and pp. 374-392.

[6]    Wade, W.R., White, J.E., and Florek, J.J., Society of Automative Engineers, Technical Paper 810118, 1981.

Helen Byrne and John Norbury
Mathematical Institute, University of Oxford
24-29 St. Giles
Oxford, U.K.

D S COHEN AND T ERNEUX
# Moving boundary problems in controlled release pharmaceuticals

## Introduction

The simplest method to deliver drugs is to introduce a relatively large quantity of drug into the body by oral or intravenous means. On the average, the drug is administered at the appropriate dosage but, at certain time points, an inappropriate high level of drug can be observed. A more recent method of drug delivery uses various membrane or polymeric devices to deliver drug over an extended period of time at a constant level to the body (from several hours to several years). Of course the drug is delivered at this level whether the body needs it or not. Moreover, it is not yet clear which drugs or bioactive agents are best given by a constant rate. For some drugs, a pattern of input may be optimal. For example, a modulated delivery system controlled by external means may improve the release pattern of insulin.

All these drug delivery problems are formulated mathematically as one–phase moving boundary problems for the concentration of the drug [1–3] with time–dependent boundary conditions. Since the release rate depends on the behavior of the moving front, the main problem is to predict its time history. This problem is difficult mathematically because the moving front may have a changing time history which precludes the use of a similarity variable.

In this paper, we consider the Higuchi simple model for drug release. The model describes the diffusion of the dissolved drug initially loaded in the polymeric device. We develop an asymptotic method to determine the position of the moving front. We then apply this method to a specific problem involving a time dependent boundary condition.

## Formulation

In the early 1960's, Higuchi [4,5] formulated a simple model for the release of a drug. A drug or bioactive agent is initially immobilized in a polymer matrix. Its loading is significantly higher than the drug solubility in the matrix. In contact with a dissolution medium (e.g., water or a biological fluid), the drug diffuses through the polymer. The slow erosion of the saturated drug reservoir controls the

161

release process. Figure 1 shows the concentration of drug at time $T = 0$ and time $T > 0$. The problem is formulated by Fick equation for the concentration C of the drug

$$C_T = DC_{XX}, \; S(T) < X < L. \tag{1}$$

The boundary conditions are given by

$$C = 0 \text{ at } X = L \tag{2}$$

$$C = C_s \text{ at } X = S(T) \tag{3}$$

and

$$-DC_X = (C_s - A)S'(T) \text{ at } X = S(T). \tag{4}$$

The first boundary condition describes the perfect sink of the drug at the fixed boundary $X = L$. The second boundary condition requires that $C = C_s$ at the moving boundary. $C_s$ is defined as the solute solubility of the drug in

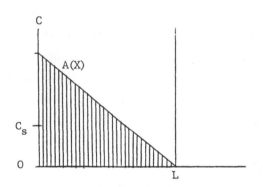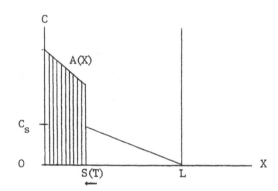

Figure 1

the polymer. The third condition represents a mass balance equation at the moving front. $A = $ constant is defined as the initial loading of the drug. The release rate is defined by

$$R = -D(C_X)_{X=L}. \tag{5}$$

The moving boundary problem can be solved exactly [6] because it has a simple $T^{1/2}$ time history. If the boundary conditions are modified and are time dependent, the position of the moving front cannot be determined. A commonly used approximation is based on the limit

162

$$\varepsilon = C_s/A \to 0 \tag{6}$$

In the next section, we consider a modification of the Higuchi problem which involves a time–dependent boundary condition and construct an asymptotic solution valid in the limit $\varepsilon \to 0$.

Variable Drug Load

An interesting modification of the Higuchi problem was proposed by Lee [7]. It consists of using a variable drug load in order to modify the $t^{1/2}$ law of the release rate. Figure 2 illustrates the new problem and represents the concentration of the drug at $T = 0$ and $T > 0$. The initial loading is given by $C = A(X)$ $(0 < X < L)$. Introducing the new variables

$$u = C/C_s, \; t = [DC_s/(A(0)L^2)]T, \; x = X/L, \; s = S/L, \tag{7}$$

into Eqs. (1) – (4), we find the following equations

$$u_t = \varepsilon^{-1}u_{xx}, \; s(t) < x < 1, \tag{8}$$

$$u = 0 \text{ at } x = 1, \tag{9}$$

$$u = 1 \text{ and } u_x = [a(x) - \varepsilon]s'(t) \text{ at } x = s(t) \tag{10}$$

where the function $a(x)$ and $\varepsilon$ are defined by

$$a(x) = A(X)/A(0) \text{ and } \varepsilon = C_s/A(0) \tag{11}$$

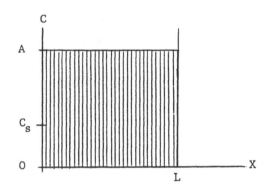 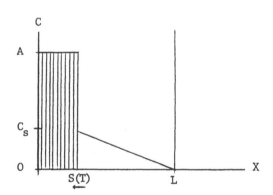

Figure 2

We now seek a solution of these equations of the form

$$u(x,t,\varepsilon) = u_0(x,t) + \varepsilon u_1(x,t) + \dots \tag{12}$$

$$s(t,\varepsilon) = s_0(t) + \varepsilon s_1(t) + \dots \tag{13}$$

163

Substituting (12) and (13) into Eqs. (8) – (10) leads to the following solution

$$u(x,t,\varepsilon) = \frac{(1-x)}{(1-s_0(t))} + 0(\varepsilon) \tag{14}$$

and

$$s(t,\varepsilon) = s_0(t) + 0(\varepsilon). \tag{15}$$

Using the second boundary condition in (10), we find that $s_0(t)$ satisfies the following equation

$$s_0'(t) = -\frac{1}{(1-s_0)a(s_0)}, \quad s_0(0) = 1. \tag{16}$$

The solution of Eq. (16) is given by the implicit relation

$$t = \int_1^{s_0} (r-1)a(r)dr. \tag{17}$$

If $a = 1$, we find that

$$s_0(t) = 1 - (2t)^{1/2} \tag{18}$$

as suggested by the exact solution for $A = $ constant. If $A(X) \neq$ constant, a different time history is possible. Eq. (16) can be studied for various $a(x)$ or we can determine $a(x)$ to obtain specific time histories. Different control mechanisms are also possible and lead to different equations for $s_0$. A detailed analysis of the effects of time–dependent boundary conditions will be given in a future publication.

References

1.  D.S. Cohen and T. Erneux, Free boundary problems in controlled release pharmaceuticals 1. Diffusion in glassy polymers. SIAM J. Appl. Math. 48, 1451–1465 (1988).

2.  D. S. Cohen and T. Erneux, Free boundary problems in controlled release pharmaceuticals 2. Swelling–controlled release. SIAM J. Appl. Math. 48, 1466 – 1474 (1988).

3.  D. S. Cohen and T. Erneux, Changing time history in moving boundary problems, SIAM J. Appl. Math. 50, 483 – 489 (1990).

4.  T. Higuchi, Rate of release of medicaments from ointment bases containing drugs in suspension, J. Pharmac. Sci. 50, 874 – 875 (1961).

5.  T. Higuchi, Mechanism of sustained–action medication. Theoretical analysis of rate of release of solid drugs dispersed in solid matrices, J. Pharmac. Sci. 52, 1145 – 1149 (1963).

6.    P. I. Lee, Effect of non—uniform initial drug concentration distribution on the kinetics of drug release from glassy hydrogel matrices, Polymer <u>25</u>, 973 − 978 (1984).

7.    P. I. Lee, Diffusional release of a solute from a polymeric matrix— approximate analytical solutions, J. Membrane Sci. <u>7</u>, 255 − 275 (1980).

**Donald S. Cohen**             and    **Thomas Erneux**
Department of Applied Mathematics       Dept. of Engineering Sciences
California Institute of Technology,      and Applied Mathematics
Pasadena, California 91125               Northwestern University,
                                         Evanston, Illinois 60208

P DE MOTTONI AND M SCHATZMAN

# Geometrical evolution of developed interfaces

**Abstract.** *Consider the reaction-diffusion equation in $\mathbb{R}^N \times \mathbb{R}^+ : u_t - h^2\Delta u + \varphi(u) = 0$; $\varphi$ is the derivative of a bistable potential with wells of equal depth and $h$ is a small parameter. If the initial data has an interface, we give an asymptotic expansion of arbitrarily high order and error estimates valid up to time $O(h^{-2})$. At lowest order, the interface evolves normally, with a velocity proportional to the mean curvature.*

## 1. Orientation

Let $\Phi(u)$ be an even smooth function, having the shape given in Fig. 1, and let $\phi = \Phi'$. The exact condition satisfied by $\Phi$ and $\phi$ are specified as follows:

$$(1.1) \qquad \begin{cases} \phi(u) < 0,\ \forall u \in (0,1), \quad \phi(u) > 0,\ \forall u \in (1, +\infty), \\ \phi'(1) > 0, \\ r\phi''(r) > 0,\ \forall r \neq 0. \end{cases}$$

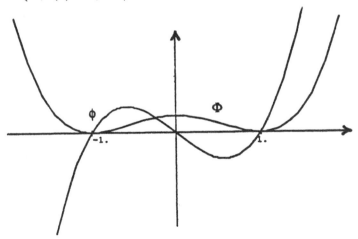

*Fig. 1. The functions $\Phi$ and $\phi$*

Let $h > 0$ be a small parameter; consider the semilinear parabolic equation in $\mathbb{R}^N \times \mathbb{R}^+$, $N > 1$

$$(1.2) \qquad \frac{\partial u}{\partial t} - h^2\Delta u + \phi(u) = 0$$

with initial condition

$$(1.3) \qquad u(x,0) = u_0(x), \quad x \in \mathbb{R}^N.$$

166

Our problem arises naturally when studying transition between phases which are equally probable, and one is not interested in the microscopic properties of the interface. Equation (1.2) can be derived from the time-dependent real Ginzburg-Landau model, and our $u$ can be understood as an order parameter.

In first place, we show that if $u_0$ is in $L^{\infty}(\mathbb{R}^N)$ (1.2), (1.3) possesses a unique solution, and we give estimates for large times of all the derivatives of $u$.

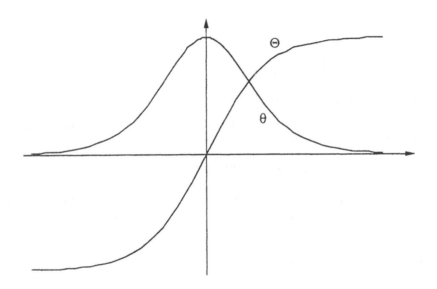

*Fig. 1.2. The standing wave $\Theta$ and its derivative $\theta$*

For not too large times, the solution of (1.2) with smooth initial data $u_0$ behaves as if there were no diffusion. In particular, it tends to $\pm 1$, according to the sign of the initial data. We choose initial data which are bounded away from zero for large $|x|$. When the gradient of $u$ is large enough (namely, $|\nabla u| = O(1/h)$), the diffusion balances the kinetic effects. The solution of (1.2),(1.3) stays smooth because of the diffusion, but it develops transition zones between the regions where $u \sim 1$ and $u \sim -1$, in which $|\nabla u|$ is very large. A dimensional argument shows that the thickness of these transition zones should be $O(h)$.

We are interested in the evolution of developped interfaces for large but not infinite times, "developped" meaning that the initial condition has an interface. The asymptotic for $t \to +\infty$ is not interesting: if $\liminf_{|x|\to\infty} u_0(x) > 0$, then it can be shown that $\lim_{t\to+\infty} u(x,t) = +1$, and symmetrically with respect to $u = 0$.

The phenomena we consider are on the time scale $h^{-2}$, in contrast with the one-dimensional case, where the relevant time scale is $\exp(C/h)$. Nevertheless, the simple aspects of the one-dimensional case are the basic building blocks for our analysis.

It is well known that the problem

(1.4)
$$\frac{\partial u}{\partial t} - \frac{\partial^2 u}{\partial x^2} + \phi(u) = 0, \quad x \in \mathbb{R}$$

admits travelling wave solutions $v(x - ct)$ which are standing, i.e., $c = 0$. This holds because $\Phi$ has minima of equal depth. Let $\Theta$ be the unique increasing solution of

(1.5)
$$-\Theta'' + \phi(\Theta) = 0$$

such that $\Theta(0) = 0$, and let $\theta = \Theta'$. Clearly, $\Theta(x)$ converges exponentially to $\pm 1$ as $x \to \pm\infty$. More precisely, there exist positive constants $\beta$ and $\beta'$ such that:

(1.6)
$$\begin{cases} 1 - \Theta(x) \sim \beta' \exp(-\beta x), \text{ for } x \to +\infty \\ \theta(x) \sim \beta\beta' \exp(-\beta x), \text{ for } x \to +\infty \\ \theta'(x) \sim -\beta^2\beta' \exp(-\beta x), \text{ for } x \to +\infty, \end{cases}$$

with similar expressions as $x$ tends to $-\infty$.

The linearization of (1.4) around $\Theta$ suggests that we consider the unbounded operator $A$ in $L^2(\mathbb{R})$ defined by by

(1.7)
$$D(A) = H^2(\mathbb{R}); \quad Au = -u'' + \phi'(\Theta)u.$$

which is clearly selfadjoint because $\phi'(\Theta)$ is real and bounded. An essential property of $A$ is summarized in the following statements:

(1.8)
$$\begin{cases} \textit{The infimum of the spectrum of } A \textit{ is the isolated simple eigenvalue } 0; \\ \textit{the corresponding eigenspace is spanned by } \theta. \end{cases}$$

Indeed, if we differentiate (1.5), we can see that $\theta$ satisfies

(1.9)
$$-\theta'' + \phi'(\Theta)\theta = 0,$$

i. e., $\theta$ is in the kernel of $A$. As $\phi'(1) = \max_x \phi'(\Theta(x))$, we know that the intersection of the spectrum of $A$ with $(-\infty, \phi'(1))$ contains only eigenvalues. Since $\theta$ does not vanish, 0 is a simple eigenvalue.

That zero is the lowest eigenvalue of $A$ is a geometrical fact, which expresses the invariance of (1.4) with respect to space translations.

With the operator $A$ we shall associate the Rayleigh quotient

(1.10)
$$\bar{R}^0[y] = \frac{\int \left(|\nabla y|^2 + \phi'(\Theta)y^2\right) dx}{\int y^2 dx}$$

defined for $y \in H^1(\mathbb{R}^N), y \neq 0$. Here, and throughout the paper, the integrals – if not otherwise specified – extend to the whole space. Clearly,

$$\inf\{\bar{R}^0[y] \ / \ y \in H^1(\mathbb{R}^N)\} = 0.$$

It is well known that interfaces appear in the one-dimensional case; it can be shown that they develop in the N-dimensional case as well.

The first guess for an approximate solution of (1.2) is

$$(1.11) \qquad u = \Theta\left(\frac{\Lambda^0(x, h^2 t)}{h}\right),$$

where $\Lambda^0(x, s)$ is the algebraic distance from $x \in \mathbb{R}^N$ to a certain hypersurface $\Gamma^0(h^2 t)$ (namely, $\Lambda^0$ is the Euclidean distance on one side of $\Gamma^0$ and minus the Euclidean distance on the other side); the slow time scale $s = h^2 t$ will be justified by the following considerations: If we apply $\partial/\partial t - h^2 \Delta + \phi$ to (1.11) we obtain

$$(1.12) \qquad \begin{aligned} &\left(\frac{\partial}{\partial t} - h^2 \Delta\right) u + \phi(u) \\ &= h\theta\left(\frac{\Lambda^0(x, h^2 t)}{h}\right) \frac{\partial}{\partial s} \Lambda^0(x, h^2 t) - \Theta''\left(\frac{\Lambda^0(x, h^2 t)}{h}\right) \left|\nabla \Lambda^0(x, h^2 t)\right|^2 \\ &\quad - h\theta\left(\frac{\Lambda^0(x, h^2 t)}{h}\right) \Delta \Lambda^0(x, h^2 t) + \phi(u). \end{aligned}$$

If $\Gamma^0(s)$ is smooth, the distance to $\Gamma^0(s)$ is smooth in a tubular neighbourhood

$$\mathcal{V}^0(s) = \{x \ / \ \mathrm{dist}(x, \Gamma^0(s)) \le \bar{\lambda}\},$$

where $\mathrm{dist}(x, A)$ is the Euclidean distance from $x$ to a set $A$, and

$$(1.13) \qquad \left|\nabla \Lambda^0(x, s)\right| = 1 \quad \text{in } \mathcal{V}^0(s)$$

Therefore (1.12) becomes by virtue of (1.5)

$$(1.14) \qquad \begin{aligned} &\left(\frac{\partial}{\partial t} - h^2 \Delta\right) u + \phi(u) \\ &= h\theta\left(\frac{\Lambda^0(x, h^2 t)}{h}\right) \left(\frac{\partial \Lambda^0}{\partial s} - \Delta \Lambda^0\right)(x, h^2 t). \end{aligned}$$

The function $\theta(x)$ takes its maximum at $x = 0$. To make the right hand side of (1.12) as small as possible, it is necessary to have

$$(1.15) \qquad \left(\frac{\partial \Lambda^0}{\partial s} - \Delta \Lambda^0\right)(x, s) = 0, \quad x \in \Gamma^0(s).$$

Relations (1.13) and (1.15) together govern the evolution of $\Gamma^0(s)$.

As we shall see below, the lowest order term (1.15) for the evolution of the interface is what geometers call motion by mean curvature: the velocity of the hypersurface is normal and proportional to the mean curvature. In other words, the global picture is essentially a dynamical Plateau Problem.

The evolution problem for $\Gamma^0(s)$, specified by equations (1.13), (1.15) is known to have a unique local solution, i.e., up to $s < T_{\text{sing}}$, where $T_{\text{sing}} > 0$.

Our approach to the problem consists in providing a sound mathematical foundation to a Chapman-Enskog expansion, corresponding to the intuition often put forward by physicists, that the leading term in the expansion should describe the evolution of the interface by motion along the gradient of the surface tension.

In this article, we give an asymptotic expansion of arbitrary large order for the solution of (1.2) with initial condition of the form

$$(1.16) \qquad u_0(x) = \Theta\left(\frac{\Lambda^0(x,0)}{h}\right)$$

in the neighbourhood of $\Gamma_0 = \{\Lambda^0(x,0) = 0\}$, which is supposed to be a $N-1$-dimensional compact manifold. The expansion is valid for times $t \leq h^{-2}T^*$, where $T^*$ is any positive number less than $T_{\text{sing}}$. The asymptotic expansion of $u$ is of the form

$$(1.17) \qquad u(x,t;h) \sim \sum_{j \geq 0} v^j\left(\frac{\Lambda(x,h^2t;h)}{h}, \Sigma(x,h^2t;h), t, h^2t\right) h^j,$$

and the asymptotic expansion of the distance function to the interface is

$$\Lambda(x,s;h) \sim \sum_{j \geq 0} \Lambda^j(x,s)h^j$$

Here $\Sigma(x,h^2t;h)$ describes tangential coordinates.

The expansion technique is fairly obvious: for all $j$ we equate to zero the coefficients of $h^j$ in the expansion of the equation. The lowest order term $v^0$ will be equal to $\Theta(\mu)$. The second term $v^1$ stays bounded for all time if we require an orthogonality condition. This condition is exactly (1.15). It is a fortunate circumstance that $v^1$ vanishes, but

$v^2$ does not, as long as $\Gamma^0(s)$ is not flat. So, in contrast to the one-dimensional case, the exact solution cannot be closer to $v^0$ than an $O(h^2)$ in $L^\infty$ norm. At higher order $(j \geq 1)$, each $v^j$ satisfies an equation of the form

$$\frac{\partial v^j}{\partial t} + Av^j = -P^j$$

with a source term $P^j$ depending on $v^k$ and its derivatives ($k \leq j-2$), and on $\Lambda^k$ and its derivatives ($k \leq j-1$). The $v^j$'s stay bounded for all time if the source term is orthogonal for large $t$'s to the kernel of the operator $A$. This leads to an equation for $\Lambda^j$ of the form

$$\frac{\partial \Lambda^j}{\partial s} - \Delta^{\Gamma^0(s)}\Lambda^j - C(\sigma,s)\Lambda^j = Q_0^j$$

170

where $\Delta^{\Gamma^0(s)}$ is the Laplace operator of the manifold $\Gamma^0(s)$, equipped with the metric induced by $\mathbb{R}^N$, $C(\sigma, s)$ is the sum of the squares of the scalar curvatures, and $Q_0^j$ depends only on $\Lambda^k$ ($k \leq j - 2$), and $\lim_{t \to +\infty} v^k$ ($k \leq j - 2$).

All the terms in the asymptotic expansion are estimated in a number of functional spaces. In particular we show that for $j \geq 1$ the quantity $|v^j(\mu, \sigma, s, t)|$ converges exponentially to zero as $\mu$ tends to infinity, uniformly in $\sigma, s, t$. This enables us to extend the expansion (1.17) defined on a tubular neighbourhood of $\Gamma^0(s)$ as a $C^\infty$ function on $\mathbb{R}^N$. Then we estimate in $L^2$ and $L^\infty$ the remainder $\rho^{q,h} = (\partial/\partial t - h^2\Delta)u^{q,h} + \phi(u^{q,h})$ where $u^{q,h}$ is the expansion truncated at $q$ terms. We have to estimate the derivatives of any order in $s$ and $\sigma$ because we perform an inductive proof which involves more and more differentiations.

The estimates on the remainder enable to appreciate the error made when taking the truncated expansion in place of the true solution. The idea of the error estimate is the following: $v$ be the exact solution of (1.2) with initial data (1.3). The difference $w = v - u^{q,h}$ satisfies

$$\frac{\partial w}{\partial t} - h^2\Delta w + \phi'(u^{q,h}) = -\rho^{q,h} - \psi(u^{q,h}, w)w^2.$$

The remainder $\rho^{q,h}$ is small if we take $q$ large enough, and $\psi$ is bounded when its arguments are bounded. Estimates in $L^2$ do not work because of the nonlinear term on the right hand side. Therefore, we combine $L^2$ and $L^\infty$ estimates in a bootstrap procedure, using connectedness in time. We provide error bounds in $L^2$ and $L^\infty$ of the whole space. It turns out that the number of terms $q$ we need to estimate the error increases linearly with the space dimension $N$. This is why we need expansions of arbitrary order.

We can perform the $L^2$ estimates because the spectrum of the self-adjoint operator $-h^2\Delta + \phi'(u^{q,h})$ can be estimated from below by $-Ch^2$. The potential $\phi'(u^{q,h})$ is everywhere strictly positive except in a region of width $O(h)$ around the interface. Therefore, all the interesting phenomena happen very close to the interface. This rough picture can be made precise using Agmon estimates, and suitably modifying their presentation by Helffer and Sjöstrand. Hence we may reduce ourselves to a Neumann problem in a tubular neighbourhood of the interface of width $O(\sqrt{h})$. As we are interested in a lower bound, we consider only the normal modes, which leads us to a collection of Neumann problems depending on the tangential variable $\sigma$ as a parameter. We deal with them by comparing them to a $\sigma$-independent Neumann problem, and precise perturbation techniques.

The symmetry assumption we made on the potential $\Phi$ was introduced only for convenience. We believe that all the results proved here go through for more general potentials, the only essential requirement being that $\Phi$ has two wells of equal depth.


The compactness assumption on $\Gamma_0$ is technical in nature. It could be replaced by suitable uniformity assumptions.

Our analysis can be applied to the one-dimensional case: then all terms of the asymptotic expansion are zero, except the zeroth order term (the same situation, intuitively characterized by the absence of surface tension, prevails in case of a flat interface), and the error estimates are exponential.

## 2. Motion by mean curvature.

Let $\bar{\Gamma}$ be a fixed reference manifold, and let $\sigma \mapsto X^0(0, \sigma, s)$ be a diffeomorphism between $\bar{\Gamma}$ and $\Gamma^0(s)$. If $\nu(\sigma, s)$ is the exterior normal to $\Gamma^0(s)$ at $X^0(0, \sigma, s)$, we define

$$(2.1) \qquad X^0(\lambda, \sigma, s) = X^0(0, \sigma, s) + \lambda \nu(\sigma, s).$$

The maping $X^0(., ., s)$ is a diffeomorphism from $[-\bar{\lambda}, \bar{\lambda}] \times \bar{\Gamma}$ to a tubular neighbourhood $\mathcal{V}^0(s)$ of $\Gamma^0(s)$. The inverse diffeomorphism is $x \mapsto \left( \Lambda^0(x, s), \Sigma^0(x, s) \right)$. This can be expressed as

$$(2.2) \qquad x = X^0(\Lambda^0(x, s) \Sigma^0(x, s), s), \quad \forall x \in \mathcal{V}^0(s).$$

Without loss of generality, we may assume that

$$(2.3) \qquad \frac{\partial X^0}{\partial s}(0, \sigma, s) \text{ is parallel to } \nabla \Lambda^0(X^0(0, \sigma, s)), \forall \sigma, s.$$

It can be shown that this assumptions freezes the parametrization of $\Gamma^0(s)$ by $\bar{\Gamma}$, up to a reparametrization of $\bar{\Gamma}$ which does not depend on time. If we differentiate (2.2) with respect to $s$, we obtain:

$$(2.4) \qquad \begin{aligned} 0 = & \frac{\partial X^0}{\partial \lambda} \left( \Lambda^0(x, s), \Sigma^0(x, s), s \right) \frac{\partial \Lambda}{\partial s}(x, s) \\ & + \frac{\partial \Sigma^0}{\partial s}(x, s) \cdot X^0 \left( \Lambda^0(x, s), \Sigma^0(x, s), s \right) + \frac{\partial X^0}{\partial s} \left( \Lambda^0(x, s), \Sigma^0(x, s), s \right). \end{aligned}$$

Here $(\partial \Sigma^0 / \partial s) \cdot X^0$ denotes the action of the vector field $\partial \Sigma^0 / \partial s$ on $X^0$ (differentiation of $X^0$ in the direction of $\partial \Sigma^0 / \partial s$).

Relation (2.1) implies that

$$(2.5) \qquad \frac{\partial X^0}{\partial \lambda} \left( \Lambda^0(x, s), \Sigma^0(x, s), s \right) = \nu \left( \Sigma^0(x, s), s \right), \quad \forall x \in \Gamma^0(s).$$

Relation (2.4) becomes, by virtue of (2.3), (2.5) and the equation for $\Lambda^0$ (1.15):

$$(2.6) \qquad \frac{\partial X^0}{\partial s} \left( \Lambda^0(x, s), \Sigma^0(x, s), s \right) = -\nu(x, s) \Delta \Lambda^0(x, s), \quad \forall x \in \Gamma^0(s).$$

A calculation in local coordinates shows that $\Delta \Lambda^0(x, s)$ is the mean curvature of $\Gamma^0(s)$, i. e., the sum of the scalar curvatures. The sign convention is that the mean curvature of a convex hypersurface is positive. Thus (2.6) is precisely the motion by mean curvature.

**P. de Mottoni**
Dipartimento di Matematica, Università di Roma Tor Vergata
**M. Schatzman**
Équipe d'Analyse Numérique, Université de Lyon I

172

# Chemical and biological reactions II

M E BREWSTER

# Numerical bifurcation analysis of free-boundary problems

We present a study of numerical determination of critical points in a family of solutions of a free-boundary problem dependent on free parameters. The problem chosen has features typical of free-boundary problems arising in combustion. Consider the parabolic differential equation

$$u_t = \Delta u + f(u, \lambda, \alpha) \tag{1}$$

where $u$ is temperature, $f$ is the reaction rate , and $\lambda$ and $\alpha$ are free parameters. We choose

$$f(u, \lambda, \alpha) = \lambda u^\alpha H(u - 1)$$

where $H$ is the Heaviside step function. We consider steady solutions in one spatial dimension. Thus, (1) becomes

$$u'' + \lambda u^\alpha H(u - 1) = 0, \tag{2}$$

The boundary conditions

$$u'(0) = 0, \quad u(1) = 0 \tag{3}$$

are also imposed. This problem has been studied by [Alexander and Fleishman, 1982] for $\alpha = 0$ and by [Stuart, 1988] for $\alpha = 2$.

We wish to numerically determine fold points on the nontrivial solution branch. There are (at least) three ways to perform local analysis of the solution space.

(1) Eulerian Approach: We linearize problem $(2, 3)$ in fixed coordinates, taking the generalized derivative of non-smooth functions. This is the classical approach to problems of hydrodynamic stability such as the Rayleigh-Taylor instability, as discussed in [Chandrasekhar]. For a solution $(u_0, \lambda, \alpha)$ with free-boundary $\xi$ (that is, $u(\xi) = 1$) to be a critical point, it is necessary that the linearized problem

$$u_1'' + \lambda(\alpha u_0^{\alpha-1} H(\xi - x) - \frac{u_0^\alpha}{u_0'(\xi)}\delta(\xi - x))u_1 = 0, \tag{4}$$

with boundary conditions (3) have a nontrivial solution.

(2) Domain Perturbation: For small changes in the free boundary, perturb the coordinates so that the free-boundary is "frozen", as presented in [Joseph, 1973, and Sijbrand, 1979].

If $X$ is the original coordinates, $\xi$ is the frozen position of the free-boundary, and $s$ is the free-boundary perturbation, let

$$X = x + s\phi(x; \xi) \tag{5}$$

where

$$\phi(0; \xi) = \phi(1; \xi), \quad \phi(\xi; \xi) = 1. \tag{6}$$

Then equation (2) becomes

$$u'' - \frac{s\phi''}{1 + s\phi'} u' + \lambda(1 + s\phi')^2 u^\alpha H(\xi - x) = 0. \tag{7}$$

A family of functions $\phi$ having the required properties is

$$\phi_n(x; \xi) = \left\{ \begin{array}{ll} \frac{x^n}{\xi^n}, & \text{for } x < \xi \\ \frac{(1-x)^n}{(1-\xi)^n}, & \text{for } x > \xi \end{array} \right\}. \tag{8}$$

We take the simplest case $n = 1$. The $\phi'$ has a jump discontinuity, and $\phi''$ is a delta function. Thus we have artificially introduced a discontinuity in the first derivative of the solution, a type of singularity which also arises in combustion problems. Equation (7) now becomes

$$u'' - 2\eta\delta(\xi - x) + \lambda m^2 u^\alpha H(\xi - x) = 0, \tag{9}$$

where

$$m = 1 + \frac{s}{\xi}, \quad \eta = -p\bar{u}', \quad p = \frac{s}{2\xi(1 - \xi) + s(1 - 2\xi)}.$$

We use an overbar to denote the average value: $2\bar{u} = u(x^+) + u(x^-)$. For a solution $(u_0, \lambda, \alpha)$ with free-boundary $\xi$ to be a critical point, it is necessary that the linearized problem (9) with boundary conditions (3) have a nontrivial solution.

(3) Boundary-Value Approach: Because of the simple nature of our model problem, the solution can be found by shooting on the interval $[0, \xi]$ with appropriate matching conditions. The solution space is thus reduced to $(u(0), \lambda, \alpha)$. Letting $u$ be the solution

of

$$u'' + \lambda u^{\alpha} = 0, \quad \text{for } x \in [0, \xi] \tag{10}$$

with

$$u(0) = v, u'(0) = 0,$$

then the matching conditions are

$$g(v, \xi, \lambda) = \left\{ \begin{array}{c} u(\xi) - 1 \\ u'(\xi) + \frac{1}{1-\xi} \end{array} \right\} = 0. \tag{11}$$

For our numerical studies, all solutions where obtained by fixing $\xi$, and solving (11) by shooting. The differential equation (10) was solved numerically using the NAg Adams method routine. The jacobian

$$\frac{\partial g}{\partial(v, \xi)}$$

must vanish at a critical point. This jacobian was evaluated numerically by difference quotients.

For our numerical representation of the problem (9), we consider a spectral (or Galerkin) Fourier approximation of the solution. That is, we let $\hat{u}_k$ be the **true** Fourier coefficients of $u$ such that

$$u(x) = \sum_{k=0}^{\infty} \hat{u}_k \cos \pi(k + \frac{1}{2})x, \quad \text{for } x \in [0, 1]. \tag{12}$$

The boundary conditions (3) are automatically satisfied. We may then approximate the Fourier coefficients of equation (9) by

$$G_k = -\pi^2(k + \frac{1}{2})^2 \hat{u}_k - 4\eta \cos \pi(k + \frac{1}{2})\xi + \lambda m^2 (\widehat{u^{\alpha}} * \hat{H})_k = 0, \tag{13}$$

where

$$\widehat{u^{\alpha}} * \hat{H} = F_N(|F_N^{-1}\hat{u}|^{\alpha} F_N^{-1}\hat{H})), \tag{14}$$

and

$$\hat{H}_k = \frac{2}{\pi(k + \frac{1}{2})} \sin \pi(k + \frac{1}{2})\xi. \tag{15}$$

176

The discretization error occurs in the evaluation of the solution with

$$F_N^{-1} \hat{u}(x) = \sum_{k=0}^{N-1} \hat{u}_k \cos \pi(k + \frac{1}{2})x.$$

This truncated summation is used to evaluate nonlinearities from data values at $x_j = \frac{j}{N}$, for $j = 0, \ldots, 4N - 1$, and the result is returned to Fourier space with the discrete Fourier transform $F_N$. We evaluate the magnitude of the jump in the first derivative by

$$\eta = p(s, \xi) \sum_{k=0}^{N-1} \pi(k + \frac{1}{2}) \hat{u}_k \sin \pi(k + \frac{1}{2})\xi. \tag{16}$$

An additional equation is obtained from the condition that $u = 1$ at the free-boundary:

$$G_N = \left( \sum_{k=0}^{N-1} \hat{u}_k \cos \pi(k + \frac{1}{2})\xi \right) - 1. \tag{17}$$

The linearizations $1, 2$ can be respresented spectrally by linearizing $G$, given in $(13, 17)$, with respect to $(\hat{u}, \xi)$ and $(\hat{u}, s)$, respectively. In each case we obtain an $N + 1$ by $N + 1$ matrix, which we denote by

$$J_N = \begin{pmatrix} A & b \\ c^T & d \end{pmatrix}, \tag{18}$$

where $A$ is an N by N matrix. For nonzero $\alpha$, $A$ is a full matrix, but for $\alpha = s = 0$,

$$A = \text{diag}(-\pi^2(k + \frac{1}{2})^2). \tag{19}$$

Also,

$$c_k = \cos \pi(k + \frac{1}{2})\xi. \tag{20}$$

The vector $b$ and scalar $d$ are different in the two linearizations. We illustrate for the case $\alpha = 0$. In the Eulerian linearization,

$$b_k = 4\eta\pi(k + \frac{1}{2}) \sin \pi(k + \frac{1}{2})\xi + 2\lambda m^2 \cos \pi(k + \frac{1}{2})\xi + \lambda \frac{\partial m^2}{\partial \xi} \hat{H}_k,$$

$$d = -\sum_{k=0}^{N-1} \pi(k + \frac{1}{2}) \hat{u}_k \sin \pi(k + \frac{1}{2})\xi = \bar{u}'. \tag{21}$$

However, in the domain perturbation linearization,

$$b_k = -4\frac{\partial \eta}{\partial s}\cos \pi(k+\tfrac{1}{2})\xi + 2\lambda \frac{\partial m^2}{\partial s}\frac{\sin \pi(k+\tfrac{1}{2})\xi}{\pi(k+\tfrac{1}{2})}, \quad d = 0. \tag{22}$$

The determinant of the matrix $J_N$ cannot be evaluated numerically because it becomes very large, ($O(N!^2)$,) as the number of modes, $N$, becomes large. However, the Shur determinant

$$S_N = d - c^T A^{-1} b \tag{23}$$

remains bounded as $N \to \infty$. For the case $\alpha = s = 0$, $A$ is diagonal and so $A^{-1}$ is easy to evaluate. However, $A$ is, in general, a full matrix, and the calculation of the Shur determinant by a direct factorization of $A$ is expensive. If this method is to be used in a practical application, the structure of $A$ should be exploited. We are currently investigating the use of the asymptotic structure of $A$ to derive iterative methods.

The error in the calculation of $S_N$ can be estimated asymptotically. As $k \to \infty$,

$$G_k \sim -\pi^2(k+\tfrac{1}{2})^2 \hat{u}_k - 4\eta \cos \pi(k+\tfrac{1}{2})\xi,$$

so

$$\hat{u}_k \sim \frac{-4\eta \cos \pi(k+\tfrac{1}{2})\xi}{\pi^2(k+\tfrac{1}{2})^2}. \tag{24}$$

Further, in linearization (2),

$$(Ax - b)_k \sim -\pi^2(k+\tfrac{1}{2})^2 x_k - 4\frac{\partial \eta}{\partial s}\cos \pi(k+\tfrac{1}{2})\xi$$

so

$$x_k \sim -\frac{4\frac{\partial \eta}{\partial s}\cos \pi(k+\tfrac{1}{2})\xi}{\pi^2(k+\tfrac{1}{2})^2} \tag{25}$$

Hence, one may show, using summation by parts,

$$\sum_{k=N}^{\infty} c_k x_k \sim C \sum_{k=N}^{\infty} \frac{\cos^2 \pi(k+\tfrac{1}{2})\xi}{(k+\tfrac{1}{2})^2} \sim \frac{C}{2N} \quad \text{where} \quad C = -\frac{4\frac{\partial \eta}{\partial s}}{\pi^2}, \tag{26}$$

as $N \to \infty$. Hence, the discretization error in $S_N$ is $O(N^{-1})$. The estimate of the error can be used to correct the Shur determinant and an asymptotic error of $O(N^{-2})$ is then observed.

178

Similar arguments can be applied to the Eulerian linearization, but the asymptotics are more complicated because the off-diagonal elements cannot be neglected as in (25). Numerical experiments show that the Eulerian linearization gives good results provided $s = 0$, (that is, the solution is continuously differentiable,) but does not converge otherwise. The effect of spectral smoothing (see,for example, [Majda, Mcdonough and Osher, 1978]) was investigated, and found to be counterproductive. This is consistent with the error analysis (26), in which the dominant error comes from the truncation of a series of positive terms whose convergence cannot be accelerated by spectral smoothing. Numerical results are presented in the following tables.

### Numerical Data: Shur Determinant

alpha=0, xi=0.5, s=0

| Fourier modes | Eulerian | Domain Perturbation |
|---|---|---|
| 16 | -.28E-03 | .51E-01 |
| 64 | -.12E-02 | .10E-01 |
| 128 | -.25E-02 | .13E-02 |
| exact | 0 | 0 |

alpha=0, xi=0.4, s=0.1

| Fourier modes | Eulerian | Domain Perturbation |
|---|---|---|
| 16 | .14 | .15 |
| 64 | -.56E-01 | .35E-01 |
| 128 | .18 | .88E-02 |
| exact | exact | 0 |

Fold Points (±0.1 in alpha), N=16

| alpha | xi |
|---|---|
| 0.0 | .50 |
| 0.1 | .52 |
| 0.5 | .60 |
| 0.9 | .70 |
| 1.0 | --- |

# References

Alexander, R.K. and Fleishman, B.A., *Perturbation and Bifurcation in a Free Boundary Problem*. J. Diff. Eq. **45** (1982), 34–52.

Chandrasekhar, S., *Hydrodynamic and Hydromagnetic Stability*, Clarendon Press, Oxford, 1961.

Joseph, D. *Domain perturbations: the higher order theory of infinitesimal water waves*, Arch. Rational Mech. Anal., **51** (1973), 295–303.

Majda, A., McDonough, J. and Osher, S., *The Fourier Method for Nonsmooth Initial Data*, Math. Comp. **32**, (1978), 1041–1081.

Sijbrand, J.,*Bifurcation analysis for a class of problems with a free boundary*, Nonlinear Analysis, **3**, (1979), 723–753.

Stuart, A.M.,*The mathematics of porous medium combustion*, from "Nonlinear Diffusion Equations and Their Equilibrium States," L. A. Peletier, W.-M. Ni and J. Serrin, eds., Springer-Verlag, Berlin, 1988.

M. E. Brewster

Department of Mathematical Sciences

Rensselaer Polytechnic Institute, Troy, New York

A DI LIDDO
# The dead-core in a single reaction with lumped temperature

## INTRODUCTION

In this paper we analyze a mathematical model for a single chemical reaction in an isothermal catalyst. We denote by u the nondimensional concentration of the reactant and by v the nondimensional temperature which is assumed to be spatially homogeneous. We assume also that the reaction is *endothermic* which means that heat is consumed during the reaction.

The equations of the model are

$$\partial u/\partial t = \Delta u - \phi g(v)f(u) \quad \text{in } \Omega$$

$$u=1 \quad \text{on } \partial\Omega$$

$$dv/dt = k(1-v) + \lambda\phi g(v)\int_{\Omega}f(u)dx$$

$$u(x,0)=1 \quad \text{in } \Omega \qquad v(0)=v_0\geq 0$$

(1)

Here, $f(u)=u^p$ if $u\geq 0$ and $f(u)=0$ if $u<0$ while $g(v)$ is the Arrhenius function $\exp(\gamma(1-1/v))$ if $v>0$ and 0 if $v\leq 0$.

The parameters $k$, $\phi$, $\alpha$, $\gamma$, $p$ are positive while $\lambda$ is negative since the reaction is endothermic. This model was first proposed in [2] as a first approximation of the nonisothermal model in which the temperature is spatially distributed.

The exotermic case is investigated in [10] when $p\geq 1$.

The same model is studied in [6] with more general initial-boundary conditions.

If $p\geq 1$ then the solution $u(x,t)$ of (1) is strictly positive at any time t. If $0<p<1$ then $u(x,t)$ can be zero in a non empty set $D(t)$ which is called the *dead-core* at time t. In $D(t)$ no reaction takes place and therefore it would be useful to avoid the existence of these regions.

For a review of the problem of the dead-core for a single parabolic equation see [9].

Let $(\underset{\sim}{u},\underset{\sim}{v})$, $(\tilde{u},\tilde{v})$ be two pairs of "smooth" functions (that is $\underset{\sim}{u}$, $\tilde{u}$ are in $C^{2,1}(\Omega\times(0,\infty))\cap C^{1,0}(\Omega\times[0,\infty))$ and $\underset{\sim}{v}$, $\tilde{v}$ are in $C^1((0,\infty))\cap C^0([0,\infty))$). We say that $(\underset{\sim}{u},\underset{\sim}{v})$, $(\tilde{u},\tilde{v})$ are pairs of lower and upper solutions (l.u.s.) if

$$\tilde{u}_t - \Delta\tilde{u} + \phi g(\underset{\sim}{v}) f(\tilde{u}) \geq 0 \qquad \underset{\sim}{u}_t - \Delta\underset{\sim}{u} + \phi g(\tilde{v}) f(\underset{\sim}{u}) \leq 0 \qquad \text{in } \Omega \times (0,\infty)$$

$$\tilde{v}_t - k(1-\tilde{v}) - \lambda\phi g(\tilde{v}) \int_\Omega f(\underset{\sim}{u}) dx \geq 0 \qquad \text{in } (0,\infty)$$

$$\underset{\sim}{v}_t - k(1-\underset{\sim}{v}) - \lambda\phi g(\underset{\sim}{v}) \int_\Omega f(\tilde{u}) dx \leq 0 \qquad \text{in } (0,\infty)$$

$$\tilde{u} - 1 \geq 0 \qquad \underset{\sim}{u} - 1 \leq 0 \qquad \text{on } \partial\Omega \times (0,\infty) \tag{2}$$

$$\underset{\sim}{u}(x,0) \leq 1 \leq \tilde{u}(x,0) \quad \text{in } \Omega \qquad \underset{\sim}{v}(0) \leq v_0 \leq \tilde{v}(0)$$

$$\underset{\sim}{u} \leq \tilde{u} \quad \text{in } \Omega \times (0,\infty) \qquad \underset{\sim}{v} \leq \tilde{v} \quad \text{in } (0,\infty)$$

If $p \geq 1$ we can prove the existence of a regular solution of (1) using an existence-comparison theorem of Pao [7]. If $p < 1$ then we cannot use that theorem because $f(u)$ in not lipschitzian and we prove the existence of a solution using an iterative scheme similar to the one proposed in [5].

Given the pairs of "smooth" functions $(\underset{\sim}{u}_{n-1}, \underset{\sim}{v}_{n-1})$, $(\bar{u}_{n-1}, \bar{v}_{n-1})$, we denote by $(\underset{\sim}{u}_n, \underset{\sim}{v}_n)$, $(\bar{u}_n, \bar{v}_n)$ the solution of the following system:

$$\bar{u}_{n,t} - \Delta\bar{u}_n + \phi g(\underset{\sim}{v}_{n-1}) f(\bar{u}_n) = 0 \qquad \text{in } \Omega \times (0,\infty)$$

$$\underset{\sim}{u}_{n,t} - \Delta\underset{\sim}{u}_n + \phi g(\bar{v}_{n-1}) f(\underset{\sim}{u}_n) = 0 \qquad \text{in } \Omega \times (0,\infty)$$

$$\bar{v}_{n,t} - k(1-\bar{v}_n) - \lambda\phi g(\bar{v}_n) \int_\Omega f(\underset{\sim}{u}_{n-1}) dx = 0 \qquad \text{in } (0,\infty)$$

$$\underset{\sim}{v}_{n,t} - k(1-\underset{\sim}{v}_n) - \lambda\phi g(\underset{\sim}{v}_n) \int_\Omega f(\bar{u}_{n-1}) dx = 0 \qquad \text{in } (0,\infty) \tag{3}$$

$$\bar{u}_n - 1 = 0 = \underset{\sim}{u}_n - 1 \qquad \text{on } \partial\Omega \times (0,\infty)$$

$$\underset{\sim}{u}_n(x,0) = 1 = \bar{u}_n(x,0) \quad \text{in } \Omega \qquad \underset{\sim}{v}_n(0) = v_0 = \bar{v}_n(0)$$

If $u_0 \in C^2(\Omega)$ then a unique smooth solution of (3) exists (see Amann [1]).

Let $u_0 \in C^2(\Omega)$ and $(\underset{\sim}{u}, \underset{\sim}{v})$ $(\tilde{u}, \tilde{v})$ be pairs of l.u.s.. The sequences $(\underset{\sim}{u}_n, \underset{\sim}{v}_n)_{n \in \mathbb{N}}$, $(\bar{u}_n, \bar{v}_n)_{n \in \mathbb{N}}$ obtained from the previous scheme starting from $(\underset{\sim}{u}_1, \underset{\sim}{v}_1) = (\underset{\sim}{u}, \underset{\sim}{v})$, $(\bar{u}_1, \bar{v}_1) = (\tilde{u}, \tilde{v})$ converge monotonically to a unique regular solution $(u,v)$ of (1) such that

$$\underset{\sim}{u} \leq u \leq \tilde{u} \quad \text{in } \Omega \times [0,\infty) \qquad \underset{\sim}{v} \leq v \leq \tilde{v} \quad \text{in } [0,\infty).$$

For the proof see [6].

In the rest of this paper we assume $p < 1$.

## THE STEADY-STATE PROBLEM

In this section we study the steady-state system

$$\Delta u - \phi g(v) f(u) = 0 \quad \text{in } \Omega$$

$$u = 1 \quad \text{on } \partial\Omega \tag{4}$$

$$k(1-v) + \lambda\phi g(v)\int_\Omega f(u)dx = 0$$

We can define pairs of l.u.s. for (4) as for (1) dropping the time dependent terms. An existence and comparison theorem can be proved in the same way. The uniqueness follows from Theorem 3.3 in [6].

**Lemma 1**
*Let $(u,v)$ the solution of (4) and $v*$ the unique positive solution of the equation $k(v*-1)=\lambda\phi|\Omega|g(v*)$. $|\Omega|$ is the measure of $\Omega$. Then*

$$0 \leq u \leq 1 \quad \text{in } \bar{\Omega} \quad \text{and} \quad 0 < v* \leq v < 1$$

*Proof.* $(0,v*)$, $(1,1)$ are pairs of l.u.s.. ■

**Lemma 2**
*Given $0<\phi_1<\phi_2$, let $(u_i,v_i)$ $i=1,2$ be the corresponding solutions of (4). Then $u_2 \leq u_1$ in $\Omega$, $v_2 \leq v_1$ and*

$$\phi_1 g(v_1) < \phi_2 g(v_2)$$

*Proof.* See [6]. ■

Define $\delta^2(\phi)=\phi g(v)$. Lemma 2 implies that the function $\delta^2(\phi)$ is increasing and then it has a limit L as $\phi$ tends to infinity. Since $v*$ defined in Lemma 1 tends to zero as $\phi$ tends to infinity, we have $\lim_{\phi\to\infty}\phi g(v*)=\lim_{\phi\to\infty}(k/|\lambda||\Omega|)(1-v*)=k/|\lambda||\Omega|$ and, since $v \geq v*$, it is

$$L \geq k/|\lambda||\Omega| \tag{5}$$

Now we give some results for (4) in the cases of simple geometry for $\Omega$.

If $\Omega$ is the half-line $[0,\infty)$, problem (4) becomes

$$u'' - \delta^2 f(u) = 0 \quad \text{in } \Omega$$

$$u(0)=1, \quad \lim_{x\to\infty} u(x) = \lim_{x\to\infty} u'(x) = 0 \tag{6}$$

$$k(1-v) + \lambda\phi g(v)\int_0^\infty f(u)dx = 0$$

The solution of (6) can be written explicitly in terms of x and $\delta$ as

183

$$u(x)=[1-x/x^*]_+^{2/(1-p)}$$

where $x^*=[\sqrt{2(1+p)}/(1-p)]/\delta$. Then $u(x)=0$ for any $x \geq x^*$.

Since $\int_0^\infty f(u)dx=\sqrt{2/(1+p)}/\delta$, we have that v must solve the equation $h(v)=(\lambda/k)\sqrt{2\phi/(1+p)}$, with $h(v)=(v-1)/\sqrt{g(v)}$. h is an increasing function of v, $h(1)=0$ and h tends to $-\infty$ as v tends to $0^+$. Therefore $v(\phi)$ is decreasing and tends to zero as $\phi$ tends to infinity. It follows that

$$\delta^2(\phi) \to (k^2/\lambda^2)[(1+p)/2]=[(1+p)/(1-p)](k/(|\lambda||\Omega_\infty|)$$

where $|\Omega_\infty|=[2/(1-p)](|\lambda|/k)$ is the limit of $x^*$ as $\phi$ tends to infinity. If we put p=0 then we obtain the equality in (5) and so that lower bound cannot be improved.

Let $\Omega=[-a,a]$. If $a \geq x^*$ then we obtain the solution of (4) matching the solutions of (6) on $[-a,\infty)$ and $(-\infty,a]$. This solution vanishes on the dead-core $\{x/|x| \leq a-x^*\}$. If $a<x^*$, then it can be seen by uniqueness that u is strictly positive in $[-a,a]$. Let's note that if $a<|\Omega_\infty|=[2/(1-p)](|\lambda|/k)$ than no dead-core occurs for any $\phi$.

If $\Omega=B_\rho(0)$ (that is the ball in $R^N$ with center O and radius $\rho$), then the solution of (6) is (see [3]) $(r/\rho)^{2/(1-p)}$ when $\delta=\delta^*=A/\rho$. A is a constant depending only on p and N. This solution vanishes only at the origin. If $\delta<\delta^*$ the solution of (4) is strictly positive while if $\delta>\delta^*$ it has a dead-core with positive measure. Since $\int_{B_\rho(0)} f(u)dx=B\rho^N$ (B depends only on p and N), we have that, in order to have $\delta=\delta^*$, it must be $\phi=\phi^*$ where

$$\phi^*=[1/g(1+C\lambda\rho^{N-2}/k)](A^2/\rho^2)$$

and $C=A^2B$.

We distinguish two cases:

-If $|\lambda|/k \geq 1/(C\rho^{N-2})$, then $\phi^*=\delta^*=\infty$ and no dead-core occurs for any $\phi$.

-If $|\lambda|/k<1/(C\rho^{N-2})$, then, since from Lemma 2 $\delta$ is increasing, we have that no dead-core occurs when $\phi<\phi^* \cdot$ and a dead-core of positive measure arises when $\phi>\phi^*$.

The previous results for a slab and a ball cannot be used to obtain estimates on the dead-core in the case of general domains because a comparison theorem between domains is not avalaible for the system (4).

If $\Omega$ is a general "smooth" bounded domain we know from [3] that a value $\underline{\delta}$ exists such that the solution of the equation

$$\Delta u-\underline{\delta}f(u)=0 \quad \text{in } \Omega, \qquad u=1 \quad \text{on } \delta\Omega$$

has a dead core of positive measure for $\delta > \underline{\delta}$ and is strictly positive in $\Omega$ for $\delta < \underline{\delta}$. (5) implies that if $k/(|\lambda||\Omega|)$ is large enough then (4) has a dead core of positive measure for large $\phi$.

## THE EVOLUTION PROBLEM

In which follows the solution of (1) will be denoted by $(u,v)$.

**Lemma 3**
*Define* $\theta = max(0, 1+\lambda\phi|\Omega|exp(\gamma)/k)$ *Then*
$$0 \leq u \leq 1 \quad in \; \bar{\Omega}x[0,\infty)$$
*and*
$$\theta(1-exp(-kt)) \leq v(t) \leq 1-(1-v_0)exp(-kt) \quad in \; [0,\infty)$$

*Proof.* $(0, \theta(1-exp(-kt)))$, $(1, 1-(1-v_0)exp(-kt))$ are pairs of l.u.s. ∎

Note that this lemma implies that $(u,v)$ is nonnegative.

Now we compare $(u,v)$ with the solution of the steady-state problem.

**Lemma 4**
*Let* $(u_\infty, v_\infty)$ *the solution of the steady-state problem*
$$\Delta u_\infty - \phi g(v_\infty)f(u_\infty)=0 \quad in \; \Omega \qquad u=1 \quad on \; \partial\Omega \tag{7}$$
$$k(1-v_\infty)+\lambda\phi g(v_\infty)\int_\Omega f(u_\infty)dx=0$$
*If* $v_0 \leq v_\infty$ *then*
$$u_\infty(x) \leq u(x,t) \leq 1 \quad in \; \Omega x[0,\infty) \qquad 0 \leq v(t) \leq v_\infty \quad in \; [0,\infty)$$

*Proof.* $(u_\infty, 0)$, $(1, v_\infty)$ are pairs of l.u.s. for (1). ∎

Denote by $D(t)$, $D_\infty$ the dead-core of the problems (1), (7) respectively. From the previous lemma it follows that if $v_0 \leq v_\infty$ then $D(t) \subset D_\infty$.

Now we compare the solution of (1) with the initial data and prove the convergence to the steady-state.

**Theorem 5**
*If* $k(1-v_0)+\lambda\phi g(v_0)|\Omega| \geq 0$ *then*
$$0 \leq u(x,t) \leq 1, \qquad v_0 \leq v(t) \leq 1 \quad \forall \; (x,t)\in\Omega x(0,\infty) \tag{8}$$

*Moreover u(x,.) is nonincreasing, v is nondecreasing and (u,v) converge to the solution of the steady-state problem as t→∞ .*
*Proof.* Since $(0,v_0)$, $(1,1)$ are pairs of l.u.s. we have (8). Define $u^*(x,t)=u(x,t+h)$, $v^*(t)=v(t+h)$ with h>0. $(u^*,v_0)$, $(1,v^*)$ are pairs of l.u.s. because of (8). The rest of the theorem follows from Sattinger [8]. ∎

From [4] and Theorem 5 we have:

$$u(x,t)>u(x,t+h) \qquad D(t+h)\subset D(t) \qquad d(\partial D(t+h),D(t))>0$$

for any t>0 and $x\in\Omega\backslash D(t+h)$.

Let $(u^*,v^*)$ be the solution of the system

$$du^*/dt=-\phi g(v^*)f(u^*)$$

$$dv^*/dt=k(1-v^*)+\lambda\phi|\Omega|g(v^*)f(u^*)$$

$$u^*(0)=u_0^*>0 \qquad\qquad v^*(0)=v_0^*\geq 0$$

It can be shown (see [6]) that $\int_0^\infty g(v^*(s))ds=+\infty$. Since

$$u^*(t)=[(u_0^*)^{1-p}-(1-p)\phi\int_0^t g(v(s))ds]_+^{1/(1-p)}$$

we have that a time T exists such that $u^*(t)=0$ for any $t\geq T$.
If $u_0^*=1$, $v_0^*=v_0$ then $(u^*,0)$, $(1,v^*)$ is a pair of l.u.s. for (1) and so

$$u^*(t)\leq u(x,t)\leq 1 \qquad 0\leq v(t)\leq v^*(t) \qquad \text{for } (x,t)\in\Omega\times(0,\infty)$$

It follows that u(x,t)>0 for any t<T.
Let $\tau=\inf\{t:D(t)\neq\emptyset\}$. If $v_0\leq 1$ then it can be proved (see [6]) that

$$\tau+\exp(-k\tau)/k\geq 1/k+\exp(-\gamma)/[(1-p)\phi]$$

**References.**

[1] H. Amann, *Fixed point equations and nonlinear eigenvalue problems in ordered Banach spaces,* SIAM Review, **18** (1976).
[2] R. Aris, *The Mathematical Theory of Diffusion and Reaction in Permeable Catalysts,* Vol. I and II, Clarendon Press, Oxford, (1975).
[3] C. Bandle, R. P. Sperb & I. Stakgold, *Diffusion and reaction with monotone kinetics,* Nonlinear Anal. TMA, **8** (1984), 321-333.

[4] C. Bandle & I. Stakgold, *The formation of the dead-core in parabolic reaction-diffusion equations*, Trans. Amer. Math. Soc., **286** (1984), 275-293.

[5] J. I. Diaz & I. Stakgold, *Mathematical aspects of the combustion of a solid by a distributed, isothermal, gas reaction*, to appear

[6] A. Di Liddo, L. Maddalena, *Mathematical analysis of a chemical reaction with lumped temperature and strong absorption*, J. Math. Anal. Appl., to appear

[7] C. V. Pao, *On nonlinear reaction-diffusion systems*, J. Math. Anal. Appl., **87** (1982), 165-198.

[8] D. H. Sattinger, *Monotone methods in nonlinear elliptic and parabolic boundary value problems*, Indiana Univ. Math. J., **21** (1972), 979-1000.

[9] I. Stakgold, *Partial extinction in reaction-diffusion*, Conferenze del Seminario di Matematica dell'Università di Bari, **224** (1987)

[10] J. M. Vega, *Invariant regions and global asymptotic stability in an isothermal catalyst*, SIAM J. Math. Anal. Appl., **19** (1988), 774-796.

Andrea Di Liddo

Istituto per Ricerche di Matematica Applicata del CNR

c/o Dipartimento di Matematica, via G. Fortunato, 70125 Bari, Italy

M A HERRERO AND J J L VELAZQUEZ

# Asymptotics near an extinction point for some semilinear heat equations

Consider the Cauchy problem

(1) $\quad u_t - u_{xx} + u^p = 0 \quad ; \quad x \in \mathbb{R} \quad , \quad t > 0$

(2) $\quad u(x,0) = u_o(x) \quad\quad ; \quad x \in \mathbb{R}$

where

(3) $\quad 0 < p < 1 \quad ,$

(4) $\quad u_o(x)$ is continuous, nonnegative, and bounded, with a single maximum at $x = 0$ and such that $u_o(-x) = u_o(x)$ for any $x$ , $\lim\limits_{x \to \infty} u_o(x) = 0$ .

Assumptions (4) are made here for convenience, but our results still hold true under less restrictive hypotheses on $u_o(x)$ .

By standard results there exists a unique nonnegative solution $u(x,t)$ of (1)–(4) . On the other hand, it is well known that such solution has some features which are absent in the superlinear case $p \geq 1$ . For instance

i) $u(x,t)$ vanishes in finite time, i.e , there exists $T > 0$ such that $u(x,t) \neq 0$ if $t < T$ and $u(x,t) \equiv 0$ for $t \geq T$ . Furthermore, if we define the extinction set $E = \{x: \text{there exists } \{x_n\} \to x, \{t_n\} \to T \text{ with } u(x_n, t_n) > 0$ for any $n \}$ , we have that $E = \{0\}$ .

ii) $u(x,t)$ develops interfaces for any $t \in (0,T)$ , even when $u_0(x)$ is positive everywhere. More precisely, there exists a continuous curve $\zeta(t)$ such that $\lim_{t \to T} \zeta(t) = 0$ and $\Omega_+(t) = \{ x : u(x,t) > 0\} = \{ x : -\zeta(t) < x < \zeta(t) \}$

cf [K], [EK], [FH], [CMM].

In [FH] a homogeneous Dirichlet problem for (1.1) in a bounded domain was considered under suitable assumptions on the initial value $u_0$ . It was shown that in such case the following estimates hold

(5) $\qquad \lim_{t \to T} (T - t)^{-\frac{1}{1-p}} u(x,t) = (1-p)^{\frac{1}{1-p}}$ , uniformly for

$|x| \leq C (T - t)^{1/2}$ with $C > 0$ ,

(6) $\qquad$ There exists constants $c_1$ and $c_2$ such that

$$\{x: |x| \leq c_1 (T - t)^{1/2} \} \subset \Omega_+(t) \subset \{ x: |x| \leq c_2 (T - t)^{1/4} \}$$

In this communication we shall describe some asymptotic results which improve (5) and (6) . Namely, we have.

THEOREM .— Let $u(x,t)$ be the solution of (1) , (2) under assumptions (3) and (4) , and let $\pm \zeta(t)$ be its interface curves. Then

$$(7) \quad \lim_{t \to T} (T-t)^{-\frac{1}{1-p}} \cdot u(\xi(T-t)^{1/2} \, |\ln(T-t)|^{1/2}, t)$$

$$= (1-p)^{\frac{1}{1-p}} \left(1 - \frac{(1-p)}{4p} \, \xi^2\right)_+^{\frac{1}{1-p}}$$

where $s_+$ = Max$\{s,0\}$ , uniformly on sets

$$|\xi| \leq C \, (T-t)^{1/2} \, |\ln(T-t)|^{1/2} \qquad \text{with} \quad C > 0 \, ,$$

$$(8) \qquad \lim_{t \to T} \frac{\zeta(t)}{(T-t)^{1/2} |\ln(T-t)|} = \left[\frac{4p}{1-p}\right]^{1/2} \, .$$

Let us make a few remarks on the proof of the Theorem. To begin with, when (1) is replaced by

$$(9) \qquad u_t - u_{xx} = u^p \quad ; \, x \in \mathbb{R} \, , \, t > 0 \, , \, p > 1$$

it is known that solutions may blow-up in finite time $T^*$ , in the sense that limsup$_{t \to T^*}$ (sup$_{x \in \mathbb{R}}$ $u(x,t)$) = $+\infty$ for some $T^* < +\infty$ . It has been already noticed in [FH] that, while extinction and blow-up are very different phenomena, they allow to be analyzed in a surprisingly similar way (for instance, (7) is very much alike to one of the main results in [GK]). As a matter of fact, a large part of the proof of the Theorem above can be made by slightly modifying the arguments in [HV1], where problem (9),

(2) was considered under assumption (4) . It is to be noticed, however, that interfaces are absent in the superlinear case, whereas they do appear in (1)–(4) and need to be analyzed by adequate techniques.

To derive (7) and (8) , a crucial role is played by the choice of a suitable functional framework. Let us define new variables as follows

$$\phi(y,\tau) = (T-t)^{-\frac{1}{1-p}} u(x,t); \; y = x(T-t)^{-1/2} , \; \tau = - \ln(T-t)$$

Then $\phi$ solves

$$\phi_\tau = \phi_{yy} - \frac{y \, \phi_y}{2} + \frac{\phi}{1-p} - \phi^p \; ; \; y \in \mathbb{R} \; , \; - \ln(t) < \tau < \infty$$

Notice that $\overline{u}(t) = (1-p)^{\frac{1}{1-p}} (T-t)^{\frac{1}{1-p}}$ is an explicit solution of

(1) which corresponds to the constant solution $\phi^* = (1-p)^{\frac{1}{1-p}}$ .

We now linearize about $\phi^*$ by setting

$$\phi = (1-p)^{\frac{1}{1-p}} + \psi$$

in which case $\psi$ satisfies

(10a)  $\psi_\tau = \psi_{yy} - \frac{y}{2} \psi_y + \psi + f(\psi)$

where

(10b)  $f(\psi) = - ((1-p)^{\frac{1}{1-p}} + \psi)^p + (1-p)^{\frac{p}{1-p}} + \frac{p \, \psi}{1-p}$

We also define $G(y,\tau)$ by

$$G = \phi^{1-p} - (1-p)$$

so that G solves

$$(11) \qquad G_\tau = G_{yy} - \frac{y}{2} G_y + G + \frac{p}{1-p} \cdot \frac{G_y^2}{(1-p) + G}$$

For $1 \leq q < \infty$ and any integer $k \geq 1$, we now define the spaces

$$L_w^q(\mathbb{R}) = \{ g \in L_{loc}^q(\mathbb{R}) : \int_\mathbb{R} |g(s)|^q e^{-s^2/4} ds < \infty \} ,$$

$$H_w^k(\mathbb{R}) = \{ g \in L_{loc}^2(\mathbb{R}) : \text{for any } j \in [0,k] ,$$

$$g^{(j)} \in L_{loc}^2(\mathbb{R}) \text{ and } \int_\mathbb{R} (g^j(s))^2 e^{-s^2/4} ds < \infty \} .$$

It is readily seen that $L_w^2(\mathbb{R})$ (resp. $L_w^q(\mathbb{R})$, $1 \leq q < \infty$, $q \neq 2$) is a Hilbert space (resp. a Banach space) when endowed with the norm

$$\|g\|_{2,w}^2 \equiv \langle g,g \rangle$$

$$= \int_\mathbb{R} (g(s))^2 e^{-s^2/4} ds \ ( \text{ resp. } \|g\|_{q,w}^q = \int_\mathbb{R} |g(s)|^q e^{-s^2/4} ds \ )$$

For any $k \geq 1$, $H_w^k(\mathbb{R})$ can be given a structure of Hilbert space in a straightforward way.

A crucial point in our approach consists in considering (10),

(11) as dynamical systems in $L_w^2(\mathbb{R})$. For instance, (10) can be written in the form

$$\Psi_\tau = A\Psi + f(\Psi)$$

where

$$A\Psi = \Psi_{yy} - \frac{y}{2} \Psi_y + \Psi \quad ; \ D(A) = H_w^2(\mathbb{R})$$

A is then a self–adjoint operator in $L^2_w(\mathbb{R})$ , having eigenvalues
$\lambda_n = 1 - \dfrac{n}{2}$ ; $n = 0, 1, 2, \ldots$ with normalized eigenfunctions
$H_n(y) = c_n \tilde{H}_n(y)$ where $c_n = (2^{n/2} (4\pi)^{1/4} (n!)^{1/2})^{-1}$ , and $\tilde{H}_n(y)$
is the standard $n^{th}$– Hermite polinomial , so that $\|H_n\|_{2,w} = 1$

for any n . It is then natural to write $\Psi(y,\tau) = \displaystyle\sum_{k=0}^{\infty} a_k(\tau) H_k(y)$ ,
and a major part of the proof consists in showing that
$\Psi(y,\tau) \simeq a_2(\tau) H_2(y)$ as $\tau \to \infty$ , and evaluating asymptotically
$a_2(\tau)$ for large $\tau$ . The convergence stated in (7) is then
obtained by a careful analysis of (11), where the

non–homogeneous term $\left[ \dfrac{(G_y)^2}{(1-p) + G} \right]$ is shown to admit suitable
bounds, so that it can be considered as a small perturbation of
the linear part along suitable sets as $t \to T$ (cf. (7)).

The results stated in the previous Theorem were formally
obtained in [GHV] by the method of matched asymptotic
expansions. The reader is referred to [HV2] for the details of
the proofs, as well as for other related results.

## REFERENCES

[CMM]   X. Chen, H. Matano and M. Mimura, Finite–point
        extinction and continuity of interfaces in a nonlinear
        diffusion equation with strong absorption, to appear.

193

[EK]    L. C. Evans and B. F. Knerr, Instanataneous shrinking of
        the support of nonnegative solutions to certain
        nonlinear parabolic equations and variational
        inequalities, Illinois J. Math. 23 (1979), 153 − 166.

[FH]    A. Friedman and M. A. Herrero, Extinction properties of
        semilinear heat equations with strong absorption , J.
        Math. Anal. and Appl. 124, 2 (1987), 530 − 546 .

[GHV]   V. A. Galaktionov, M. A. Herrero and J.J.L. Velázquez,
        The structure of solutions near an extinction point in a
        semilinear heat equation with strong absorption: a
        formal approach, to appear

[GK]    Y. Giga and R. V. Kohn, Asymptotically self similar
        blow-up of semilinear heat equations, Comm. Pure and
        Appl. Math. 38 (1985) , 297 - 319 .

[HV1]   M. A. Herrero and J. J. L. Velázquez, Blow-up behaviour
        of one-dimensional semilinear parabolic equations, to
        appear,

[HV2]   M. A. Herrero and J. J. L. Velázquez , Approaching an
        extinction point in one-dimensional semilinear heat
        equations with strong absorption, to appear.

[K]     A. S. Kalashnikov, The propagation of disturbances in
        problems of nonlinear heat conduction with absorption,
        USSR Comp. Math. and Math. Phys. 14 (1974) , 70 - 85 .
M. A. Herrero and J. J. L. Velázquez
Departamento de Matemática Aplicada
Facultad de Matemáticas Universidad Complutense
28040  Madrid, Spain
194

L RUBINSTEIN AND L RUBINSTEIN
# On dynamic theory of osmosis

Osmotically induced mass transfer through semipermeable membranes is an important process in biology in general, and in cytology, physiology and botany in particular. In spite of this, and in spite of the fast development of theoretical biology, osmosis, treated as a non–steady process, is not on the list of processes studied mathematically. This is even more true concerning the free boundary problems, of which the osmotically induced transport through deformable membranes represents a virtually unexplored example, worth of study. Moreover, which is more important, the dynamics of osmosis is not well understood [16],[1],[13],[14]. In particular, it is not clear whether the one–phase approach could suffice to describe the dynamics of osmosis or the use of multiphase description is imperative. In [12], Silberberg suggested on the basis of non–equilibrium thermodynamics a set of relations which was expected to form a skeleton for a consistent one–phase description of the dynamics of osmosis. Unfortunately, no initial–boundary value problems were considered in [12], so that, based on it, a conclusive decision could hardly be made concerning the suitability of the one–phase formalism to describe an actual dynamic (time dependent) osmotic process (e.g., membrane deformation,swelling or shrinking)

In this paper we attempt at making a choice between the one–phase and two–phase approaches to the dynamics of osmosis by analyzing a series of easily tractable one–dymensional models of increasing complexity. Each next level in this hierarchy is introduced to mimic some dynamic osmotic feature, inexplicable on the previous level of description. All these models concern a thick semipermeable membrane, bounding a vessel (osmometer) filled with water solution of a diffusive component (impermeant), unable to penetrate the membrane. The solid elements of the membrane are modelled as elastic springs in a viscous surrounding.

We begin with the analysis of the one–phase diffusion model which could be viewed as the simplest possible version of the one–phase description of the kind employed in [12]. This model (as all other models analyzed) employs the assumptions of local mechanical and chemical equilibrium both in the bulk phases and at the membrane boundaries; incompressibility of components is assumed.

The molar concentrations $c_1(x,t)$ and $c_2(x,t)$ of water in the osmometer and in the membrane respectively are assumed to evolve by the simple Fick's diffusion in the system of average volume velocity [8] Across the free boundaries $x = X_1(t)$ , $x = X_2(t)$ (osmometer/membrane and membrane/external water basin interfaces, respectively) the mechanical and chemical equilibrium is assumed along with Stefan diffusion conditions for all components in the system. Besides, equations of state, expressing the incompressibility of mixture components and linear dependence of diffusion fluxes of solution components inside each compartment are used. By using zero divergency of the average volume velocity in a mixture of incompressible liquids and continuity of its normal components at the surfaces of a strong discontinuity [4], one may prove that water concentration and its diffusion flux are continuous across the free boundary $x = X_1(t)$. This, together with continuity of stresses and chemical potential of water at $x = X_1(t)$, allows to prove that if

$$Z(t) \overset{def}{=} t^{1/2}\frac{d}{dt}X_1(t) < \infty, \quad \tilde{U}_1(t) = t^{1/2}\frac{d}{dt}c_1^1[X_1(t),t] \tag{I.1_1}$$

$$w_1(t) = t^{1/2}\frac{\partial}{\partial x}c_1^1[X_1(t),t], \quad c_1^0 \overset{def}{=} \underset{t \downarrow 0}{\text{Lim}}\, c_1^1[X_1(t),t] \tag{I.1_2}$$

then

196

$$\operatorname*{Lim}_{t \downarrow 0} w_1(t) = 0 \land c_1^0 \neq c_{10} \overset{\text{def}}{=} c_1^1(x,0) \equiv \text{const} \tag{I.2}$$

On the other hand, one may prove, using the heat potential technique and the same Stefan condition, that

$$w_1(t) = \pi^{-1/2}(c_{10} - c_1^0)f(t) + W_1(t \mid \tilde{U}_1, w_1, Z) \tag{I.3}$$

where

$$\operatorname*{Lim}_{t \downarrow 0} f(t) = 1, \ \operatorname*{Lim}_{t \downarrow 0} W_1 = 0 \tag{I.4}$$

so that

$$\operatorname*{Lim}_{t \downarrow 0} w_1(t) \neq 0 \tag{I.5}$$

in contradiction with (I.2). The contradiction obtained proves that the solution of the problem does not exists, i.e. that the employed one–phase model is inconsistent.

One proceeds next to the analysis of the following three two–phase capillary–diffusion models. The first two models are of auxiliary nature and concern, respectively: a) a rigid capillary membrane bounding an osmometer, open to the outer space through a manometer tube, and b) an elastic capillary membrane of constant thickness (i.e. incapable of swelling or shrinking) sealing the osmometer. For these two models the global existence, uniqueness and stability theorems are proved, and it is shown that the corresponding osmotical systems tend asymptotically to their thermodynamic and mechanical equilibrium.

Finally, the third model concerns an elastic membrane, capable of swelling and shrinking, sealing the osmometer. The corresponding periodicity element of the system, consisting of 3 spatial domains (osmometer $G_1$, the membrane $G_3 \cup G_4 \cup G_5$, external basin $G_2$, respectively) is represented schematically in Fig.1. The membrane $G_3 \cup G_4 \cup G_5$ is viewed as consisting of the capillary $G_3$, separated from the membrane body $G_4$ by another rigid capillary wall membrane $G_5$. The membrane body is viewed as essentially another osmometer, containing thus an osmotically active impermeable component along with the permeant (water) and the elastic (spring) element.

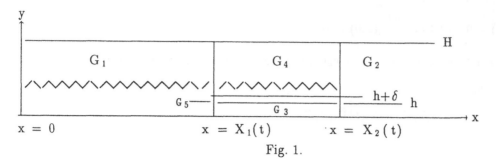

Fig. 1.

The notations are

$$G = \overset{5}{\underset{1}{U}} G_i = \overset{5}{\underset{1}{U}} (D_{it} \times G^i) \tag{II.1.1$_1$}$$

where

$$G^k = \{y: 0 < y < H\}\ k = 1,2,\ G^3 = \{y: 0 < y < h\} \tag{II.1$_2$}$$

$$G^4 = \{y: h+\delta < y < H\},\ G^5 = \{y: h < y < h + \delta\} \tag{II.1$_3$}$$

and

$$D_{1t} = \{x: 0 < x < (X_1(t)\},\ D_{2t} = \{x: X_2(t) < x < \infty\} \tag{II.1$_4$}$$

$$D_{kt} = \{x: (X_1(t) < x < X_2(t)\}\ \ k = 3,4,5 \tag{II.1$_5$}$$

$$L(t) = X_2(t) - X_1(t) \tag{II.2$_1$}$$

so that $L(t)$ is the membrane thickness at a moment t. We assume that

$$\delta \ll h \ll H,\ L(t) \ll X_1(t)\ \ \ \forall\, t \geq 0 \tag{II.2$_2$}$$

Besides the unknown positions of the membrane interfaces $X_1(t)$, $X_2(t)$, the dependent variables are $c_j^i$ —water and impermeant molar concentrations , $p_i$ — pressure and $u_i$ —water velocity in each domain $G_i$ ($i = 1,...,5$, $j = 0,1$ for $i = 1$, $j = 1$ for $i = 2,3,5$, $j = 1,2$ for $i = 4$) All inertia terms in the momentum equations for each component are neglected (the local mechanical equilibrium). The local chemical equilibrium is assumed everywhere including the interfaces. A linear pressure drop is assumed in the wall membrane $G_5$. All processes are reduced to one–dimensional ones by means of averaging all governing equations in the y direction. In order to obtain correct conditions for mass transfer through the membrane boundaries (i.e. through the boundaries of m.m.b and through the capillary $G_3$) the equations of continuity are averaged over the respective

198

(two–dimensional) regions $(G_i, i = 1,2,...,5)$.

The governing equations thus are: equation of diffusion of the impermeant in $D_{1t}$; the equation of convection–diffusion of water in $D_{4t}$, including the source term, describing the water influx into $D_{4t}$ from $D_{3t}$; the Darcy's law for the water motion in $D_{3t}$ and $D_{4t}$, the quasistationary equation for the spring motion in $D_{4t}$, taking into account the friction with water in $D_{4t}$ and $D_{3t}$; the condition of local chemical equilibrium (continuity of the chemical potential of water at the boundaries $x = X_i(t)$ of $D_{3t}$ ; the conditions of local mechanical equilibrium at the boundaries $x = X_i(t)$, $i = 1,2$. These conditions are complemented by the initial conditions: constant initial concentrations in all compartments and zero initial deformation of the springs.

The model suggested is thus essentially of a two–phase nature — water in it may be in one of two states — free, flowing in the capillary phase $D_3$, or bound in the membrane matrix $D_4$

The crucial step of our analysis concerns the initial direction of evolution. It is shown that, depending on the choice of physical parameters, the following cases may occur:

$P_1$: *The process begins with swelling of both the osmometer and of the membrane*

$$\frac{d}{dt}X_1(0) > 0 \ \wedge \ \frac{d}{dt}L(0) > 0 \tag{II.3$_1$}$$

$P_2$: *The process begins with shrinking of both the osmometer and of the membrane*

$$\frac{d}{dt}X_1(0) < 0 \ \wedge \ \frac{d}{dt}L(0) < 0 \tag{II.3$_2$}$$

$P_3$: *The process begins with swelling of the osmometer and shrinking of the membrane*

$$\frac{d}{dt}X_1(0) > 0 \ \wedge \ \frac{d}{dt}L(0) < 0 \tag{II.3$_3$}$$

$P_4$: *The process begins with shrinking of the osmometer and swelling of the membrane*

$$\frac{d}{dt}X_1(0) < 0 \ \wedge \ \frac{d}{dt}L(0) > 0 \tag{II.3$_4$}$$

$P_5$: *At the initial moment*

$$\frac{d}{dt} \cdot X_1(0) = 0 \ \wedge \ \frac{d}{dt} \cdot L(0) = 0 \tag{II.3$_5$}$$

$P_6$: *At the initial moment*

$$\frac{d}{dt} \cdot X_1(0) > 0 \ \wedge \ \frac{d}{dt} \cdot L(0) = 0 \tag{II.3$_6$}$$

$P_7$: *At the initial moment*

$$\frac{d}{dt} \cdot X_1(0) < 0 \ \wedge \ \frac{d}{dt} \cdot L(0) = 0 \tag{II.3$_7$}$$

$P_8$: *At the initial moment*

$$\frac{d}{dt} \cdot X_1(0) = 0 \ \wedge \ \frac{d}{dt} \cdot L(0) > 0 \tag{II.3$_8$}$$

$P_9$: *At the initial moment*

$$\frac{d}{dt} \cdot X_1(0) = 0 \ \wedge \ \frac{d}{dt} \cdot L(0) > 0 \tag{II.3$_9$}$$

*Thus the two–phase capillary–diffusion model is indeed able to predict swelling or shrinking of the osmometer and of the membrane*

A more detailed analysis is required for determination of the initial direction of evolution in the indefinite cases $P_{5-9}$ or in the cases $P_{3-4}$, when the direction of evolution is definite, but where the initial velocities of water in $D_{3t}$ and $D_{4t}$ and the initial flux of water from the capillary to m.m.b. are not monotonous functions of x.

Using the results obtained and referring to the contraction mapping arguments one may prove the local existence, uniqueness ans stability theorems. Existence of a global solution (proved by the method of continuation) is a corollary of monotonicity of $X_1(t)$, which is proved by referring to the maximum principle and to the Hopf lemma.

Three possibilities are to be examined:

$$A: X_1^e \stackrel{def}{=} \lim_{t \uparrow \infty} X_1(t) < \infty \ , \ L^e \stackrel{def}{=} \lim_{t \uparrow \infty} L(t) < \infty \tag{II.4}$$

$$B: X_1^e = \infty \ , \ L^e = \infty \tag{II.5}$$

$$C: X_1^e < \infty \ , \ L^e = \infty \tag{II.6}$$

It is proved that the options B and C are impossible so that the case A is valid.

*The main overall conclusion of our analysis is that the system under consideration tends asymptotically for $t \uparrow \infty$ to its mechanical and thermodynamic equilibrium.*

200

## REFERENCES

1. Dainty J.,1963, Water relations of plant Cells, Advances Bot.Res.1,279–324
2. Enikeeva E.,1968, On the crystallization of a binary alloy with eutectic. Latvian State Annual,4,122–147.
3. Katchalsky A., Curran F.,1967, Nonequilibriuum thermodynamics, Harvard books in biophysics 1, Harvard University Press, MA.
4. Kotchin N.E.,1926, Sur la theory des ondes de chock dans les fluids; Rend. Cir.Mat.Palermo, 50,305
5. Loewy Ariel G, Siekevitz Philip, 1969 Cell structure and functions,Holt,Rinehalt and Wilson, Second edition.
6. Rubinstein L. 1956, On determining the location of boundary of separation of two slightly compressible fluids, percolating through deformable porous medium., Proc. Ufa Oil Institute 1, 75–109, Ufa, UFNI (Russian).
7. Rubinstein L. 1971, The Stefan Problem, Translation of Math. Monogr. 27, AMS Providence.
8. Rubinstein L.,1974, Passive transfer of low–molecular non–electrolytes across deformable membranes. I.Equations of convective–diffusion transfer of non–eLectrolytes across deformable membranes of a large curvature; Bull.Math.Biol.36,4,365–377.
9. Rubinstein L.,1980, Application of the integral equation technique to the solution of several Stefan problems. In. Magenes E. (ed) Free Boundary Problems, Proc. Sem. held in Pavia Sept.– Oct.1979 1, 399–416, Roma.
10. Rubinstein L. 1972, Temperature fields in oil strata, Nedra,Moscow (Russian)
11. Rubinstein L.   1982, Global stability of the Neumann solution of the two–phase Stefan problem,IMA J.Ap.Mth.,28.287–299
12 Silberberg A .,1989, Transport through deformable matrices,Biorheology 26,291–313
13. Suourivajan, S. 1970, Reverse osmosis,Lagos,London.
14. Thain J.F.,1967, Principles of Osmotic Phenomena, The Royal Institute of Chemistry
15. Тихонов А.,1938, О функциональных уравнениях типа Volterra и их приложениях к некоторым задачам математической физики, Бюллетень М.Г.У.,Секция А, Математика и Механика,1,1 – 25.
16. Хвольсон О.Д. 1892–1915, Курс физики, 2. (German translation: Chwolson O.D., 1918, Lehrbuch der Physik, Bd 1 Abt.2, Lehre von den Gasförming Körpern.

L.Rubinstein[0]& L.Rubinstein[00]

[0] Blaustein Institute for Desert Research and Department of Mathematics, Ben Gurion University of the Negev, Sede Boqer Campus, Israel.
[00] School of Applied Science and Technology, The Hebrew University of Jerusalem, Israel.

L SANTOS

# Diffusion with gradient constraint and evolutive Dirichlet condition

**Abstract**

We consider a diffusion problem with gradient constraint and evolutive Dirichlet condition. We prove, under certain assumptions on the data, its equivalence with a free boundary problem and study the existence, uniqueness and regularity of the solution and the asymptotic behavior of the solution and free boundary when $t \longrightarrow +\infty$.

Let us consider the equation

$$u_t + \nabla . \vec{q} = \beta, \tag{1}$$

denoting $u$, for instance, a density of population, a concentration of a chemical compound or a temperature, where $\vec{q} = -\lambda \nabla u$, $\lambda \in k(|\nabla u|)$ is a nonlinear diffusion coefficient and $k$ the maximal monotone graph

$$k(s) = \begin{cases} 1 & \text{if } s < 1, \\ [1, +\infty[ & \text{if } s = 1. \end{cases} \tag{2}$$

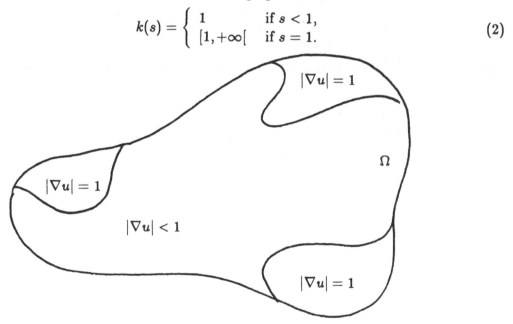

We have the following situation:

- in the region $\{|\nabla u(t)| < 1\}$, $u$ obeys the Fick or Fourier type law $u_t - \Delta u = \beta$,

- the saturation zone, where the gradient attains the threshold 1, corresponds to a free boundary, $\partial\{|\nabla u(t)| = 1\} \cap \partial\{|\nabla u(t)| < 1\}$.

We approximate the maximal monotone graph $k$ by the exponentialy type regular functions, as we can see in the following figure

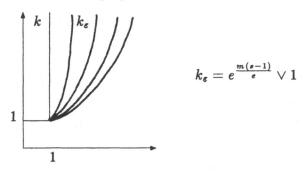

$$k_\epsilon = e^{\frac{m(s-1)}{\epsilon}} \vee 1$$

and so, our problem can be interpreted as the limit of a exponential type laws (see [6], [8]).

Let $\Omega$ be an open regular subset of $I\!\!R^n$, $\Omega_0 = \Omega \times \{0\}$, $I = ]0, T[$, $Q_T = \Omega \times I$, $\Sigma = \partial\Omega \times I$ and $\beta = \beta(t)$, $t \in I$.

Consider the problem

$$\begin{cases} u_t - \nabla.(\lambda\nabla u) = \beta \text{ in } Q_T, \\ \lambda \in k(|\nabla u|^2), \quad |\nabla u(t)| \leq 1 \text{ for a.e. } t \in I, \\ u = g \text{ on } \Omega_0 \cup \Sigma. \end{cases} \tag{3}$$

**Remark 1** *If $g \geq 0$ and $\beta \geq 0$, then $u \geq 0$. The problem with $g \equiv 0$ on $\Sigma$ was considered in [8]* $\qquad\square$

Let us present a simple example: consider $\Omega = ]0, 1[$ and the Dirichlet boundary condition $g(0, t) = 0$ and $g(1, t) = 1$. We can see that $(x, c)$ are solutions of problem (3), for all $c \geq 1$ and so, $\lambda$ need not to be unique.

**Theorem 2** [9] *Suppose that the Dirichlet condition $g$ verifies*

$$g \in L^\infty(0, T; C^2(\overline{\Omega})), \quad g_t \in L^\infty(Q_T), \tag{4}$$

$\exists r, \ 0 \leq r < 1 : \quad |g(x, t) - g(y, t)| \leq r\bar{d}(x, y) \quad \forall x, y \in \partial\Omega, \text{ for a.e. } t \in I,$

*denoting $\bar{d}$ the geodesic distance in $\Omega$.*

*Then problem (3) has a solution $(u, \lambda)$. Furthermore, $u$ is unique and, if $g$ is independent of $x$, $\lambda$ is also unique.* $\qquad\square$

**Remark 3** *The geodesic distance in $\Omega$ is defined as follows:*

$$\bar{d}(x, y) = \inf_\gamma L(\gamma), \quad \forall x, y \in \Omega,$$

*where $\gamma$ is any poliginal line contined in $\Omega$ joining $x$ to $y$. (see [2]).* □

Denoting sup=$\bigvee$ and inf=$\bigwedge$, let

$$\overline{\varphi}(x,t) = \bigvee\{v(x) : v \in \mathbb{K}_\nabla(t)\} = \min_{z \in \partial\Omega}\{g(z,t) + \overline{d}(x,z)\}, \tag{5}$$

$$\underline{\varphi}(x,t) = \bigwedge\{v(x) : v \in \mathbb{K}_\nabla(t)\} = \max_{z \in \partial\Omega}\{g(z,t) - \overline{d}(x,z)\}, \tag{6}$$

where

$$\mathbb{K}_\nabla(t) = \{v \in H^1\Omega : |\nabla v| \leq 1,\ v_{|\partial\Omega} = g(t)_{|\partial\Omega}\}.$$

If $(u, \lambda)$ is solution of problem (3), then $u$ is solution of the following equivalent variational problems (see [3], [5] or [7])

$$\begin{cases} u(0) = g(0), \\ u(t) \in \mathbb{K}_\nabla(t) \text{ for a.e. } t \in I, \\ \displaystyle\int_\Omega u_t(t)(v - u(t)) + \int_\Omega \nabla u(t).\nabla(v - u(t)) \geq \beta(t)\int_\Omega (v - u(t)), \\ \qquad\qquad\qquad \forall v \in \mathbb{K}_\nabla(t), \text{ for a.e. } t \in I, \end{cases} \tag{7}$$

$$\begin{cases} u(0) = g(0), \\ u(t) \in \mathbb{K}(t) = \{v \in H^1\Omega : \underline{\varphi}(t) \leq v \leq \overline{\varphi}(t),\ v_{|\partial\Omega} = g(t)_{|\partial\Omega}\} \text{ for a.e. } t \in I, \\ \displaystyle\int_\Omega u_t(t)(v - u(t)) + \int_\Omega \nabla u(t).\nabla(v - u(t)) \geq \beta(t)\int_\Omega (v - u(t)), \\ \qquad\qquad\qquad \forall v \in \mathbb{K}(t), \text{ for a.e. } t \in I, \end{cases} \tag{8}$$

**Proposition 4** *If $g$ satisfies condition (4), the solution $u$ of problem (3) belongs to* $W_p^{2,1}(Q_T) = L^p(0,T;W^{2,p}(\Omega) \cap W^{1,p}(0,T;L^p(\Omega)),\ 1 \leq p < +\infty.$ □

Consider now the limit elliptic problem

$$\begin{cases} u_\infty \in \mathbb{K} = \{v \in H^1(\Omega) : \underline{\varphi}_\infty \leq v \leq \overline{\varphi}_\infty,\ v_{|\partial\Omega} = g_\infty\}, \\ \displaystyle\int_\Omega \nabla u_\infty.\nabla(v - u_\infty) \geq \beta_\infty\int_\Omega (v - u_\infty),\ \forall v \in \mathbb{K}, \end{cases} \tag{9}$$

where $\overline{\varphi}_\infty(x) = \min_{z \in \partial\Omega}\{g_\infty(z) + \overline{d}(x,z)\}$ and $\underline{\varphi}_\infty(x) = \max_{z \in \partial\Omega}\{g_\infty(z) - \overline{d}(x,z)\}$.

**Remark 5** *When the boundary condition $g_\infty$ is zero, this problem is the well-known elastic-plastic torsion problem ([2], [5], [6], [7].*

Concerning to the asymptotic behavior of $u(t)$, we have the following

**Theorem 6** ([8]) *Suppose g satisfies condition (4). If*

$$\int_t^{t+1} |\beta(\tau) - \beta_\infty|^2 \xrightarrow[t \to +\infty]{} 0 \quad and \quad \int_t^{t+1} \int_{\partial\Omega} |g(\tau) - g_\infty)|^2 \xrightarrow[t \to +\infty]{} 0,$$

*then*

$$u(t) \xrightarrow[t \to +\infty]{} u_\infty \ in \ C^{0,\alpha}(\overline{\Omega}), \quad 0 < \alpha < 1.$$

*If, furthermore, $g_t \in L^\infty(0, \infty; L^1(\partial\Omega))$,*

$$\exists k_1 > 0 \quad \int_t^{t+1} \left\{ \int_\Omega |\beta'|^2 + \int_{\partial\Omega} |g_{tt}| \right\} \le k_1 \quad \forall t > 0,$$

*then,*

$$u(t) \xrightarrow[t \to +\infty]{} u_\infty \ in \ W^{1,p}(\Omega), \quad \forall p, \ 1 \le p < +\infty.$$

$\square$

Let

$$S(t) = \{x \in \Omega : |\nabla u(x,t)| = 1\}, \quad S_\infty = \{x \in \Omega : |\nabla u_\infty(x)| = 1\},$$

$$I^+(t) = \{x \in \Omega : u(x,t) = \overline{\varphi}(x,t)\}, \quad I^-(t) = \{x \in \Omega : u(x,t) = \underline{\varphi}(x,t)\},$$

$$I_\infty^+ = \{x \in \Omega : u_\infty(x) = \overline{\varphi}_\infty(x)\}, \quad I_\infty^-(t) = \{x \in \Omega : u_\infty(x) = \underline{\varphi}_\infty(x)\}.$$

**Proposition 7** (see [2], [9])

$$S(t) = I^+(t) \cup I^-(t), \quad S_\infty = I_\infty^+ \cup I_\infty^-,$$

*except in a null set.*

$\square$

Extending the results of [8], we have the following asymptotic behavior in time of the free boundaries $\partial\{|\nabla u(t)| = 1\} \cap \Omega$, when $t \longrightarrow +\infty$:

**Theorem 8** *Let $\chi(t)$ and $\chi_\infty$ be the characteristic functions of the sets $S(t)$ and $S_\infty$ respectively. Suppose that*

$$\underline{\varphi}_\infty(x) < \overline{\varphi}_\infty(x), \ \forall x \in \Omega,$$

$$\beta_\infty \ne \Delta\underline{\varphi}_\infty, \quad \beta_\infty \ne \Delta\overline{\varphi}_\infty, \quad a.e. \ in \ \Omega,$$

$$\int_t^{t+1} |\beta(\tau) - \beta_\infty|^2 d\tau \xrightarrow[t \to +\infty]{} 0, \quad \int_t^{t+1} |\beta'(\tau)|^2 d\tau \xrightarrow[t \to +\infty]{} 0,$$

*and*

$$\int_{\partial\Omega} |g_t(t)| \xrightarrow[t \to +\infty]{} 0, \quad \int_t^{t+1} \int_{\partial\Omega} |g_{tt}| \xrightarrow[t \to +\infty]{} 0.$$

*Then*

$$\chi(t) \xrightarrow[t \to +\infty]{} \chi_\infty \ in \ L^p(\Omega), \quad \forall p, \quad 1 \le p < +\infty.$$

□

**Remark 9** *Notice that this is a different free boundary problem from the oxygen comsuption problem* (see [1], [4]), *although the equivalence between* (7) *and* (8) *allow a formal analogy.*

# References

[1] C. Baiocchi and G. A. Pozzi, *An evolution variational inequality related to a diffusion-absorption problem*, Appl. Math. and Optimization **2** 4 (1976) 304-314.

[2] L. A. Caffarelli and A. Friedman, *Reinforcement problems in elasto-plasticity*, Roc. Mount. J. Math. **10** (1980) 155-184.

[3] M. Chipot, *Variational inequalities and flow in porous media*, Appl. Math. Sc. **52**, Springer-Verlag, New York, 1984.

[4] J. Crank and S. Gupta, *A moving boundary problem arising from the diffusion of oxygen in absorbing tissue*, J. Inst. Math. Appls. **10** (1972) 19-33.

[5] A. Friedman, *Variational principles and free boundary problems*, Wiley-Interscience, New York, 1982.

[6] C. Gerhardt, *On the existence and uniqueness of a warpening function in the elastic-plastic torsion of a cylindrical bar with multiply connected cross-section*, Lect. Notes in Math. (Springer) **503** (1976) 328-342.

[7] J. F. Rodrigues, *Obstacle problems in mathematical physics*, North-Holland, Amesterdam, 1987.

[8] L.Santos, *A diffusion problem with gradient constraint*, Inter. Ser: Num. Meth. (Birkhauser) **88** (1989) 389-400.

[9] L.Santos, *Strong solutions to elliptic and parabolic problems with quasilinear monotone discontinuities*, Nonlin. Anal. Theo. Meth. and Appl. Vol 17 **9** (1991) 811-824.

Lisa Santos

Departamento de Matemática

Universidade do Minho    4700 Braga, Portugal

# Control and identification

M BROKATE AND J SPREKELS
# Optimal control of shape memory alloys with solid-solid phase transitions

In this paper, we establish first order necessary optimality conditions

for an optimal control problem associated with thermomechanical

phase transitions in shape memory alloys. The mathematical model

consists of the following initial-boundary value problem: In

$\Omega_T = \Omega \times (0,T)$, $\Omega = (0,1)$, $T > 0$ , we have

(1) $\qquad u_{tt} - \left[\theta F_1'(\epsilon) + F_2'(\epsilon)\right]_x + u_{xxxx} = f(x,t)$

(2) $\qquad\qquad \theta_t - \theta F_1'(\epsilon)\epsilon_t - \theta_{xx} = g(x,t)$

(3) $\qquad\qquad\qquad \epsilon = u_x$

The initial conditions are given by $(x \in \bar{\Omega})$

(4) $\qquad u(x,0) = u_0(x)$ , $u_t(x,0) = u_1(x)$ , $\theta(x,0) = \theta_0(x)$,

and the boundary conditions are $(0 \leq t \leq T)$

(5) $\qquad u(0,t) = u_{xx}(0,t) = 0 = u(1,t) = u_{xx}(1,t)$,

(6) $\qquad \theta_x(0,t) = 0$ , $-\theta_x(1,t) = \theta(1,t) - \theta_\Gamma(t)$.

The equations (1) and (2) describe the balance of momentum and

enery in a piece of metallic alloy of unit length which, according to (5)

and (6), is simply supported at both ends and thermally insulated at x = 0 . The variables u,$\epsilon$ and $\theta$ have the usual meaning of displacement, strain and absolute temperature (as dimensionless quantities). Moreover, the time evolution is controlled by the body force f , the heat source g , and the outside temperature $\theta_\Gamma$ . A more detailed description of the physical background is given in [5], [6]. Following [2], [3] , the specific form of $F_1$ and $F_2$ for shape memory alloys is assumed to be

$$\text{(7)} \qquad F_1(\epsilon) = \epsilon^2 , F_2(\epsilon) = -\theta_c\epsilon^2 - \epsilon^4 + \epsilon^6 ,$$

where $\theta_c$ is a critical phase transition temperature. As a consequence of (7) , the stress-strain relation with the stress

$$\text{(8)} \qquad \sigma = \theta F_1'(\epsilon) + F_2'(\epsilon)$$

is not monotone and exhibits temperature dependent hysteresis.

Control problem (CP)

Minimize

$$\text{(9)} \qquad J(u,\theta; f,g,\theta_\Gamma) = \int_0^T \int_\Omega \phi_1(u_x(x,t), \theta(x,t))dxdt$$

$$+ \int_0^T \int_\Omega \phi_2(f(x,t), g(x,t))dxdt + \int_0^T \phi_3(\theta_\Gamma(t))dt$$

subject to (1) - (7) and $(f,g,\theta_\Gamma) \in K$, where $K \subset Z$ ,

$$\text{(10)} \qquad Z = H^1(0,T;H^1(\Omega)) \times H^1(0,T;H^1(\Omega)) \times H^1(0,T),$$

and $\phi_1, \phi_2, \phi_3$ are given functions.

We first summarize all regularity assumptions concerning the initial–

209

boundary value problem and the control problem.

<u>Assumption (A)</u>

(i) Regularity and compatibility of the boundary data:

$u_0 \in \tilde{H}^4(\Omega) = \{u \in H^4(\Omega) \mid u(0) = u''(0) = 0 = u(1) = u''(1)\}$

$u_1 \in H_0^1(\Omega) \cap H^2(\Omega)$ , $\theta_0 \in H^2(\Omega)$ , $\theta_0 > 0$ in $\overline{\Omega}$,

$\theta_0'(0) = 0$ , $\theta_\Gamma^0 := \theta_0'(1) + \theta_0(1) > 0$ .

(ii) Regularity of the cost functional:

$\phi_1, \phi_2 \in C^2(\mathbb{R}^2)$ , $\phi_3 \in C^1(\mathbb{R})$; $\phi_2, \phi_3$ are convex.

(iii) Regularity and compatibility of the control constraint:

$K \subset Z$ is closed, convex and bounded w.r.t. $\|\cdot\|_z$ , and for all $(f,g,\theta_\Gamma) \in K$ we have $g \geq 0$ on $\overline{\Omega}_T$ and $\theta_\Gamma(0) = \theta_\Gamma^0$ , $\theta_\Gamma > 0$ on $[0,T]$ .

◇

We now consider the observation operator $S : K \to X \times Y$

(11)                                  $(u,\theta) = S(f,g,\theta_\Gamma),$

which yields the solution of (1) - (7) corresponding to $(f,g,\theta_\Gamma)$, where

$X = W^{2,\infty}(0,T;L^2(\Omega)) \cap W^{1,\infty}(0,T;H_0^1(\Omega) \cap H^2(\Omega)) \cap L^\infty(0,T;\tilde{H}^4(\Omega))$

$Y = H^1(0,T;H^1(\Omega)) \cap L^2(0,T;H^3(\Omega)).$

We already know that $S$ is well defined:

<u>Theorem 1 ([6])</u>

If (A) holds, then (1) - (7) has a unique solution $(u,\theta) \in X \times Y$ for any $(f,g,\theta_\Gamma) \in K$ , and $S(K)$ is bounded in $X \times Y$.

210

Let now $(f,g,\theta_\Gamma) \in K$. An element $(h,k,l) \in Z$ is called admissible variation at $(f,g,\theta_\Gamma)$, if $(f,g,\theta_\Gamma) + \lambda(h,k,l) \in K$ for some $\lambda > 0$. Let us consider the following linearization of the initial-boundary value problem (1) - (7):

$$(12) \quad \varphi_{tt} + \varphi_{xxxx} = h + [F_1'(\epsilon)\psi + (\theta F_1''(\epsilon) + F_2''(\epsilon))\varphi_x]_x \quad \text{in } \Omega_T$$

$$(13) \quad \psi_t - \psi_{xx} = k + \theta F_1''(\epsilon)\epsilon_t\varphi_x + F_1'(\epsilon)\epsilon_t\psi + \theta F_1'(\epsilon)\varphi_{xt} \quad \text{in } \Omega_T$$

$$(14) \qquad\qquad \varphi(x,0) = \varphi_t(x,0) = \psi(x,0) = 0 \quad \text{in } \overline{\Omega}$$

$$(15) \qquad \varphi(0,t) = \varphi_{xx}(0,t) = 0 = \varphi(1,t) = \varphi_{xx}(1,t) \qquad \text{in } [0,T]$$

$$(16) \qquad \psi_x(0,t) = 0 \,,\, -\psi_x(1,t) = \psi(1,t) - l(t) \quad \text{in } [0,T].$$

Theorem 2

Let (A) hold. Let $(h,k,l) \in Z$ be an admissible variation at $(f,g,\theta_\Gamma) \in K$. Then $S : K \to B$ has a directional derivative $(\varphi,\psi)$ at $(f,g,\theta_\Gamma)$ in the direction $(h,k,l)$ which solves (12) - (16), where $(u,\theta) = S(f,g,\theta_\Gamma)$, $\epsilon = u_x$, and $B \supset X \times Y$,

$$B = [W^{1,\infty}(0,T);L^2(\Omega)) \cap L^\infty(0,T;H_0^1(\Omega) \cap H^2(\Omega))] \times$$

$$\times [L^2((0,T;H^1(\Omega)) \cap L^\infty(0,T;L^2(\Omega)].$$

Proof: First we show that $S$ is Lipschitz continuous on $K$ with respect to some norm intermediate between those of $B$ and $X \times Y$. The main ingredients of this proof are a test of the equations for $u = u^{(1)} - u^{(2)}$, $\theta = \theta^{(1)} - \theta^{(2)}$, by $u_t$ and $\theta$ respectively, and the inequalities of Young, Nirenberg and Cronwall. A similar procedure applied to the remainder

$$r(\lambda) = S(f + \lambda h, g + \lambda k, \theta_\Gamma + \lambda l) - S(f,g,\theta_\Gamma) - \lambda(\varphi,\psi]$$

yields the result. Due to the nonlinearities, the reasoning is not at all straightforward and somewhat lengthy; we have to refer to [1] .

◇

We now consider the control problem (CP). As it is customary in optimal control, we introduce the adjoint systems for the variables $p$ and $q$ at the optimal solution $(u^*,\theta^*) = S(f^*,g^*,h^*)$ :

(17)  $p_{tt} + p_{xxxx} - [D_1\phi_1(u_x^*,\theta^*) + (\theta^* F_1''(u_x^*) + F_2''(u_x^*))p_x +$

$\qquad + \theta_t^* F_1'(u_x^*)q + \theta^* F_1'(u_x^*)q_t]_x = 0 \quad$ in $\Omega_T$,

(18)  $q_t + q_{xx} - F_1'(u_x^*)p_x + F_1'(u_x^*)u_{xt}^* q = D_2\phi_1(u_x^*,\theta^*) \quad$ in $\Omega_T$,

(19)  $\qquad p(x,T) = p_t(x,T) = q(x,T) = 0 \quad$ in $\Omega$,

(20)  $p(0,t) = p_{xx}(0,t) = 0 \; p(1,t) = p_{xx}(1,t) \quad$ in $[0,T]$,

(21)  $\qquad q_x(0,t) = 0 = q_x(1,t) + q(1,t) \quad$ in $[0,T]$.

The main result of this paper is the following theorem.

Theorem 3

Let (A) hold, let $(u^*,\theta^*) \doteq S(f^*,g^*,\theta_\Gamma^*)$ be a solution of the control problem (CP). Then there exists a weak solution $(p^*,q^*)$ of the adjoint system (17) - (21) with $q^* \in Y$ and

$$p^* \in H^1(0,T);L^2(\Omega)) \cap L^\infty(0,T;H_0^1(\Omega) \cap H^2(\Omega)),$$

such that for any admissible variation $(h,k,l)$ at $(f^*,g^*,\theta_\Gamma^*)$ we have

212

$$\int\limits_{0}^{T} \int\limits_{\Omega} [(D_1\phi_2(f^*,g^*) - p^*)h + (D_2\phi_2(f^*,g^*) - q^*)k]\,dxdt +$$

$$+ \int\limits_{0}^{T} [\phi_3'(\theta_\Gamma^*(t)) - q^*(1,t)]l(t)dt \geq 0.$$

<u>Proof:</u> The problem is to obtain the regularity of the adjoint $(p^*,q^*)$ as stated in the theorem. As an intermediate step, we derive an a-priori-estimate (uniformly in m) of the expression

$$\sup_{t}[\|p_t^{(m)}\|^2 + \|p_{xx}^{(m)}(t)\|^2] + \int\limits_{0}^{T} \|q_{xt}^{(m)}(t)\|^2 + \|q_{xxx}^{(m)}(t)\|^2dt$$

for the Galerkin approximations $(p_t^{(m)},q_t^{(m)})$. (Here, $\|\cdot\|$ denotes the norm of $L^2(\Omega)$). For this, we use the techniques already mentioned above as well as parabolic regularity theory from [4] , applied to the equation for $q$ and to the equation for $z := q_x$. ◇

[1] Brokate,M., Sprekels, J.: Optimal control of thermomechanical phase transitions in shape memory alloy: Necessary conditions of optimality, Manuscript, submitted 1990.

[2] Falk,F., Landau theory and martensitic phase transitions, Journal de Physique C4, 12 (1982), 3-15.

[3] Falk,F., Ginzburg-Landau theory and static domain walls in shape-memory alloys, Phys. B-Condens. Matter, 54, 1983, 177-185.

[4] Ladyshenskaya,O.A., Solonnikov,V.A.,Uralceva,N.N., Linear and Quasilinear Equations of Parabolic Type, Amer. Math. Soc., Providence,R.I., 1968.

[5] Sprekels,J., Onedimensional thermomechanical phase transitions with nonconvex potentials of Ginzburg-Landau type, Preprint No. 505, Institute of Mathematics and its Applications (IMA), University of Minnesota (1989).

[6] Sprekels.J.,Zheng,S., Global solutions to the equations of a Ginzburg-Landau theory for structural phase transitions in shape memory alloys, Physica D, 39 (1989), 59-76.

Martin Brokate[1] and Jürgen Sprekels[2]


[1]FB Mathematik, Universität Kaiserslautern, D-6750, West Germany
[2]FB 10, Universität-GH Essen, D-4300 Essen 1, West Germany

M C DELFOUR* AND J-P ZOLÉSIO*

# Adjoint state in the control of variational inequalities

ABSTRACT. *In this paper we review a penalization method used to justify a Lagrangian approach to optimization problems with non-linear state equations. It is applied to an optimization problem where the state is the solution of a variational inequality. A simple example illustrates the main results and draw some comparison with classical results.*

**1. Introduction.** There are several mathematical techniques to justify the Lagrangian approach in Sensitivity Analysis for Shape Identification and Control problems. For convex cost functions and linear boundary value problems, everything can be justified by introducing a Lagrangian and making use of theorems on the differentiability of a saddle point. These basic results can be extended to semiconvex cost functionals. (cf. Correa and Seeger [1], Delfour and Zolésio [1,2,3,4]).

The Lagrangian approach is however more difficult to justify for nonlinear boundary value problems where the convexity is lost. Yet it is well known that the results are formaly the same as in the linear case. Variational inequalities provide another example which is not easily amenable to a Lagrangian formulation. In both cases a penalization method has been used to justify the Lagrangian approach (cf. Delfour and Zolésio [5, 6]). In the case of variational equations we construct an adjoint variable which is the solution of a variational inequality. Since the underlying concepts are rather technical, we give an application to the control of variational inequalitites and a one dimensional example is completely worked out. The results are presented for convex cost functional, but they readily extend to the semiconvex case.

**2. A Penalization Method for Problems with Nonlinear State Equation.** We consider state equations arising from the minimization of an energy functional $E(t, \varphi)$ over a convex closed subset $K$ of a reflexive Banach space $B$

$$E(t, y^t) = \inf_{\varphi \in B} E(t, \varphi), \qquad y^t \in B, t \geq 0. \tag{1}$$

When $K = B$ and $E(t, \varphi)$ is non-quadratic in $\varphi$ we obtain a nonlinear problem

$$y^t \in B, \quad dE(t, y^t; \varphi) = 0, \qquad \forall \varphi \in B; \tag{2}$$

* The research of this author has been supported by a Killam Fellowship from Canada Council, NSERC Grants OGP 0008730 and INF 0007939 and a FCAR Grant from the Ministère de l'Éducation du Québec.

215

when $K \neq B$ we obtain variational inequalities of the form

$$y^t \in K, \quad dE(t, y^t; \varphi - y^t) \geq 0, \qquad \forall \varphi \in K. \tag{3}$$

When $K = B$ the main difficulty in the application of the theorem of section 2 is the loss of convexity of the functional

$$\varphi \mapsto dE(t, \varphi; \psi). \tag{4}$$

For instance the convex functional

$$E(\varphi) = \frac{1}{5} \int_\Omega |\varphi|^5 \, dx - \int_\Omega \varphi \, dx \tag{5}$$

over the open domain $\Omega$ yields

$$dE(\varphi; \psi) = \int_\Omega |\varphi|^3 \varphi \psi \, dx - \int_\Omega \varphi \, dx \tag{6}$$

This functional is convex when $\varphi \geq 0$ on $\Omega$ and concave when $\varphi \leq 0$ on $\Omega$.

For variational inequalities the construction of the Lagrangian functional becomes more delicate. By adding one variable the problem.

$$J(t) = F(t, y^t), \tag{7}$$

where $y^t$ is the solution of (3) can be reformulated as an inf sup problem

$$J(t) = \inf_{\varphi \in K} \sup_{\substack{\mu \geq 0 \\ \psi \in K}} \{F(t, \varphi) - \mu \, dE(t, \varphi; \psi - \varphi)\} \tag{8}$$

but the standard results are not readily applicable here.

**2.  A penalization method.**   To get around this difficulty we use the following penalization method which can be found in Delfour and Zolésio [5,6], critical details and hypotheses can be found.

Let $E : \mathbb{R}^+ \times K \to \mathbb{R}$ be an **energy functional** defined over a closed convex subset $K$ of a Banach space $B$. For each $t$ in an interval $[0, T]$,   $T > 0$ we assume that

$$\varphi \to E(t, \varphi) \tag{9}$$

is convex and continuous on $K$ and that there exists a unique solution $y^t \in K$ to the minimization problem

$$y^t \in K, \quad E(t, y^t) = \inf_{\varphi \in K} E(t, \varphi) \stackrel{\Delta}{=} e(t) \tag{10}$$

and that $y^t$ is the unique solution of the variational inequality

$$y^t \in K, \quad dE(t, y^t; \varphi \to y^t) \geq 0, \qquad \forall \varphi \in K, \tag{11}$$

216

where

$$dE(t, \phi; \psi) = \lim_{\substack{\theta > 0 \\ \theta \to 0}} [E(t, \varphi + \theta\psi) - E(t, \varphi)]/\theta.$$

Associate with the solution $y^t$ of (10) the cost function

$$J(t) = F(t, y^t) \tag{13}$$

for some given cost functional

$$F = \mathbb{R}^+ \times K \to \mathbb{R}. \tag{14}$$

The main objective was to show that, under appropriate hypotheses, the cost function $J(t)$ can be expressed in the form

$$J(t) = J(0) + \int_0^t f(s)ds \tag{15}$$

for some function $f$ in $L^\infty(0, T)$ which will be characterized in terms of the state $y^t$ and the solution $p^t$ of an appropriate unilateral problem for each $t$. When $f$ belongs to $C^0[0, T]$, $J$ belongs to $C^1[0, T]$ and

$$dJ(t) = \lim_{\substack{s > 0 \\ s \to 0}} [J(t + s) - J(t)]/s = f(t). \tag{16}$$

The basic idea is to introduce a family of minimization problems indexed by $\varepsilon > 0$

$$J_\varepsilon(t) = \inf_{\varphi \in K} F_\epsilon(t, \varphi) \tag{17}$$

where

$$F_\varepsilon(t, \varphi) = F(t, \varphi) + \frac{1}{\varepsilon}[E(t, \varphi) - E(t, y^t)]. \tag{18}$$

Under appropriate hypotheses,

$$y_\varepsilon^t \to y^t \text{ in } B \quad \text{and} \quad J_\varepsilon(t) \to J(t) \quad \text{as} \quad \varepsilon \to 0. \tag{19}$$

So we introduce the function

$$p_\varepsilon^t = (y_\varepsilon^t - y^t)\varepsilon \tag{20}$$

and show that

$$p_\varepsilon^t \to p^t \text{ in } V \text{ (weak) as } \varepsilon \to 0, \tag{21}$$

where $V$ is Hilbert space such that $B \subset V$. Then we use the problem indexed by $e > 0$. The minimizing element $y_\varepsilon^t$ of (17) is characterized by

$$y_\varepsilon^t \in K, \quad dF(t, y_\varepsilon^t; \varphi - y_\varepsilon^t) + \frac{1}{\varepsilon} dE(t, y_\varepsilon^t; \varphi - y_\varepsilon^t) \geq 0, \quad \forall \varphi \in K. \tag{22}$$

So we can use the theorem on the derivative of a Min with respect to $t$

$$dJ_\varepsilon(t) = \partial_t F_\varepsilon(t, y_\varepsilon^t) = \partial_t[E(t, y_\varepsilon^t) + \frac{1}{\varepsilon}\partial_t[E(t, y_\varepsilon^t)E(t, y^t)]] \tag{23}$$

and the identity

$$J_\varepsilon(t) = J_\varepsilon(0) + \int_0^t dJ_\varepsilon(s)ds. \tag{24}$$

As $e$ goes to zero we obtain

$$J(t) = J(0) + \int_0^t f(s)ds, \tag{25}$$

$f \in L^\infty(0, T)$ is given by

$$f(t) = \partial_t F(t, y^t) + \partial_t dE(t, y^t; p^t). \tag{26}$$

The variables $y^t$ and $p^t$ are solution of the inequalities

$$y^t \in K, \quad dE(t, y^t; \varphi - y^t) \geq 0, \quad \forall \psi \geq \in K \tag{27}$$

and, under hypothesis (H) given below,

$$p^t \in S(t), \quad \forall \psi \in S(t), \quad dF(t, y^t; \psi - p^t) + d^2 E(t, y^t; \psi - p^t; p^t) \geq 0, \tag{28}$$

where

$$S(t) = T_K(y^t) \cap \nabla E(t, y^t)^\perp \tag{29}$$

$$T_K(y^t) \quad = V - \text{closure } \{\lambda(\varphi - y^t) : \varphi \in K, \quad \lambda \geq 0\} \tag{30}$$

$$\nabla E(t, y^t)^\perp = V - \text{closure } \{\psi \in B : dE(t, y^t; \psi) = 0\}. \tag{31}$$

Assumption $H$ is given by the identity

$$S(t) = \overline{co}A(t) \tag{H}$$

$$A(t) = \left\{ \psi \in V \left| \begin{array}{l} \exists\{\varphi_\varepsilon : \varepsilon > 0\} \subset K, \quad \psi_\varepsilon = (\varphi_\varepsilon - y^t)/\varepsilon \text{ such that} \\ \psi_\epsilon \to \psi \text{ in } V \text{ (weak) as } \varepsilon \to 0 \quad \text{and} \\ \lim_{\varepsilon \searrow 0} \frac{1}{\varepsilon} dE(t, y^t; 0, \psi_\varepsilon) = 0. \end{array} \right. \right\} \tag{32}$$

Hypothesis $(H)$ is quite abstract. Fortunately it can be easily shown that

$$\mathbb{R}^+(K - y^t) \cap \nabla E(t, y^t)^\perp \subset A(t) \subset S(t) = T_K(y^t) \cap \nabla E(t, y^t)^\perp. \tag{33}$$

So $(H)$ is always verified in finite dimension or when the so-called F. Mignot [1]'s condition is satisfied.

**3. Limiting behaviour as $t$ goes to zero.** The results from the previous section only indicate that $J(t)$ is differentiable almost everywhere. One would like to say something about the existence of $dJ(0)$. This requires the existence of the limit

$$\frac{1}{t} \int_0^t f(s)ds \to f^* \text{ as } t \to 0 \tag{34}$$

and to obtain a complete characterization of $dJ(0) = f^*$ in term of $y$ and the adjoint variable we must be able to say something about the limit points of $p^t$ as $t$ goes to zero. It turns out that for $t$ in a neighborhood of $t = 0$, the $p^t$'s are bounded in the Hilbert space $V$. So we can extract subsequences such that

$$p^{t_n} \to p^* \text{ in } V(weak). \tag{35}$$

The difficult part is to characterize $p^*$. In general $p^*$ is not the solution of

$$p \in S(0), \quad \forall \psi \in S(0), \quad dF(0, y; \psi - p) + d^2 E(0, y; \psi - p; p) \geq 0, \tag{36}$$

since the sets $S(t)$ do not "converge" to $S(0)$. In the case $K = B$, most of the above difficulties dissappear and we recover the expected results under reasonable assumptions.

**4. Application to the control of variational inequalitites.** Since the previous results are quite abstract, it is useful to specialize them to a simple example which brings out the important features. To further simplify we shall even focus our attention of a classical control problem which is much simpler than Shape problems since the control enters linearly in the variational inequality.

We need the following ingredients

$H = L^2(\Omega)$ with the usual norm $|\cdot|$ and inner product $(\cdot, \cdot)$

$V = H_0^1(\Omega)$ with the usual norm $||\cdot||$ and inner product $((\cdot, \cdot))$

$a : V \times V \to \mathbb{R}$ bilinear continous and coercive

$K = \{\varphi \in V : \varphi \geq 0 \text{ a.e. in } \Omega\}, \quad f \in H, \quad v \in H \text{ (control variable)}.$

Consider the solution $y$ of the variational inequality

$$y = y(v) \in K, \quad a(y, \varphi - y) \geq (f + v, \varphi - y), \quad \forall \psi \in K \tag{37}$$

and the associated cost function

$$J(v) = \frac{1}{2}\{|y(v) - z_d|^2 + |v|^2\}. \tag{38}$$

219

The corresponding energy and cost functionals are

$$E(t, \varphi) = \frac{1}{2}a(\varphi, \varphi) - (f + u + tv, \varphi) \tag{39}$$

$$F(t, \varphi) = \frac{1}{2}\{|\varphi - z_d|^2 + |u + tv|^2\}. \tag{40}$$

Here F. Mignot [1]'s condition holds and from the previous section

$$J(u + tv) = J(u) + \int_0^t (u + sv - p^s, v)ds \tag{41}$$

where

$$y^s \in K, \qquad a(y^s, \varphi - y^s) \geq (f + u + sv, \varphi - y^s), \quad \forall \varphi \in K \tag{42}$$

$$p^s \in S(s), \qquad a(p^s, \psi - p^s) + (y^s - z_d, \psi - p^s) \geq 0, \qquad \forall \psi \in S(s). \tag{43}$$

It can also be shown that

$$\|p^t\| \leq \frac{2}{\alpha}(t|v|^{+1}y^0 - z_d|). \tag{44}$$

So there are weak limit points of $\{p^t\}$ in $V$. Let

$$p = p(u, v) \in V \tag{45}$$

be such that

$$\exists\{p^{t_n}\}, \quad p^{t_n} \in S(t_n) \quad \text{such that} \quad p^{t_n} \to p \quad \text{in } V \text{ (weak)}. \tag{46}$$

As a result

$$a(p, p) + (y^0 - z_d, p) \leq 0. \tag{47}$$

At the minimum

$$\exists u^* \in H, \quad J(u^*) = \inf_{v \in H} J(v). \tag{48}$$

So for a fixed $v$, denote by $p(u^*, v)$ a weak limit point of $p^t = p(u^* + tv)$. Then

$$J(u^* + tv) - J(u^*) = \int_0^t (u^* + sv - p(u^* + sv), v)ds \tag{49}$$

which implies that

$$(u^* - p(u^*, v), v) \geq 0, \quad \forall v \in H. \tag{50}$$

REMARK 2.1. At this stage it is important to notice that the limit points $p(u^*, v)$ are dependent on $u^*$ and the direction $v$. This is the generalization of the adjoint variable in differentiable problems. So far we have not been able to show that $p(u^*, v)$ is unique for the class of problems under consideration. However it will be unique and completely characterized by a variational inequality in the finite dimensional example given in the next section. ◆

REMARK 2.2. The characterization of the minimum given in (50) is different from the classical one given by MIGNOT and PUEL [1] which, if interpreted correctly, gives: if $y^* = y(u^*)$ is the state corresponding to $u^*$, then

$$\exists \bar{p} \in S(0), \quad \forall \psi \in S(0), \quad a(\psi, \bar{p}) \leq (y^* - z_d, \psi) \tag{51}$$

$$\bar{p} + u^* = 0. \tag{52}$$

◆

## 5. A simple finite dimensional example. Let

$$K = \{\varphi \in \mathbb{R} : \varphi \geq 0\}, \quad E(u, \varphi) = \frac{1}{2}\varphi^2 - u\varphi \tag{53}$$

$$F(u, \varphi) = \frac{1}{2}\{(\varphi + 1)^2 + u^2\}, \quad u \in \mathbb{R} \tag{54}$$

Given $u$, the solution $y = y(u)$ of the variational inequality

$$y \geq 0, \quad (y - u)(\varphi - y) \geq 0, \quad \forall \varphi \geq 0 \tag{55}$$

is given by

$$y(u) = \begin{Bmatrix} u, & u \geq 0 \\ 0, & u < 0 \end{Bmatrix}, \tag{56}$$

and the associated cost function by

$$J(u) = F(u), y(u)) = \frac{1}{2} \begin{Bmatrix} (u+1)^2 + u^2+, & u \geq 0 \\ 1 + u^2, & u < 0 \end{Bmatrix}. \tag{57}$$

The directional semi derivative at $u$ in the direction $v$ is given by

$$dJ(u; v) = \begin{Bmatrix} (2u+1)v, & u \geq 0 \\ \max\{v, 0\}, & u = 0 \\ uv, & u \leq 0 \end{Bmatrix}. \tag{58}$$

It is readily checked that $J$ has a unique minimum at $u^* = 0$ where the function $J$ is not differentiable.

Now we apply the results from the previous section to this example at $u = 0$ for $v \in \mathbb{R}$ and $t > 0$:

$$y^t = \begin{Bmatrix} tv, & v \geq 0 \\ 0, & v < 0 \end{Bmatrix}, \quad \nabla E(t, y^t)^\perp = \begin{Bmatrix} \mathbb{R}, & v \geq 0 \\ \{0\}, & v > 0 \end{Bmatrix}. \tag{59}$$

221

$$T_K(y^t) = \left\{ \begin{array}{ll} \mathbb{R}, & v > 0 \\ \mathbb{R}^+, & v \leq 0 \end{array} \right\}, \quad S(t) = \left\{ \begin{array}{ll} \mathbb{R}, & v > 0 \\ \mathbb{R}^+, & v = 0 \\ \{0\}, & v < 0 \end{array} \right\}, \quad S(0) = \mathbb{R}^+. \tag{60}$$

By solving the variational inequality

$$p^t \in S(t), \quad (p^t, \psi - p^t) - (tv, \psi - p^t) \geq 0, \quad \forall \psi \in S(t), \tag{61}$$

we obtain

$$p^t = \left\{ \begin{array}{ll} -1, & v \geq 0 \\ 0, & v \geq 0, \end{array} \right\}, \quad p^0 = -1. \tag{62}$$

Finally

$$J(tv) = J(0) + \int_0^t (sv - p^s)v \, ds \tag{63}$$

and

$$\lim_{t \searrow 0} \frac{J(tv) - J(O)}{t} = -p^*(0, v)v, \tag{64}$$

where

$$p^*(0, v) = \lim_{s \searrow 0} \qquad p^s = \left\{ \begin{array}{ll} -1, & v \geq 0 \\ 0, & v < 0, \end{array} \right\}. \tag{65}$$

Notice that for $v < 0\, p^*(0, v) \neq p^0$ so that in general the limit points of $\{p^t\}$ as $t$ goes to zero are not the solution of (61) for $t = 0$ and that they do indeed depend on the direction $v$.

Finally at $u^* = 0$ the optimality condition (50) gives

$$-p^*(0, v)v \geq 0, \quad \forall v \in \mathbb{R}. \tag{66}$$

REMARK 2.3. It is readily seen that $p^*(0, v)$ is the unique solution of the variational inequality

$$p = p^*(0, v) \in S^*(0, v), \quad (p, \psi - p) \geq 0, \quad \forall \psi \in S^*(0, v), \tag{61}$$

where $S^*(0, v) = \mathbb{R}$, if $v > 0$, $\mathbb{R}^+$, if $v = 0$ and $\{0\}$ if $v < 0$. Moreover $S^*(0, v) = S(0)$ for $v \geq 0$ and $S^*(0, v) \neq S(0)$ for $v < 0$. ◆

REMARK 2.4. Again, if interpreted correctly, the condition (51) (52) yields

$$y^* = y(0) = 0 \tag{67}$$

$$\exists \bar{p} \in S(0) = \mathbb{R}^+, \quad \forall \psi \in S(0) = \mathbb{R}^+, \quad \psi \bar{p} \leq \psi \Rightarrow 0 \leq \bar{p} \leq 1 \tag{68}$$

and

$$\bar{p} + u^* = \bar{p} + 0 = 0 \Rightarrow \bar{p} = 0. \tag{69}$$

As indicated in Remark 3.2, our characterization (66) of the minimum is different from the one given by MIGNOT and PUEL [1]. But, up to a change in the sign of $\bar{p}$, the two conditions applied to this special example do not seem incompatible. ◆

# REFERENCES

R. CORREA AND A. SEEGER [1], *Directional derivatives of a mimimax function*, Nonlinear Analysis Theory, Methods and Applications **9** (1985), 13-22.

M.C. DELFOUR AND J.P. ZOLÉSIO [1], *Dérivation d'un MinMax et application à la dérivation par rapport au contrôle d'une observation non-différentiable de l'état*, C.R. Acad. Sc. Paris, Sér. I **302** (1986), 571-574.

[2], *Shape Sensitivity Analysis via Min Max Differentiability*, SIAM J. on Control and Optimization **26** (1988), 834-862.

[3], *Differentiability of a Min Max and Application to Optimal Control and Design Problems, Part I*, in "Control Problems for Systems Described as Partial Differential Equations and Applications," I. Lasiecka and R. Triggiani, eds., Springer-Verlag, New York, 1987, pp. 204-219.

[4], *Differentiability of a Min Max and Application to Optimal Control and Design Problems, Part II*, in "Control Problems for Systems Described as Partial Differential Equations and Applications," I. Lasiecka and R. Triggiani, eds., Springer-Verlag, New York, 1987, pp. 220-229.

[5], *Further Developments in Shape Sensitivity Analysis via a Penalization Method*, in "Boundary Control and Boundary Variations," J.P. Zolésio, ed., Springer–Verlag, Berlin, Heidelberg, New-York, Tokyo, 1988, pp. 153-191.

[6], *Shape Sensitivity Analysis via a Penalization Method*, Annali di Matematica Pura ed Applicata **CLI** (1988), 179-212.

F. MIGNOT [1], *Contrôle dans les inéquations variationnellles elliptiques*, J. Funct. Anal. **22** (1976), 130-1857.

F. MIGNOT AND J.P. PUEL [1], *Optimal Control in some Variational Inequalities*, SIAM J. Control and Optimization **22** (1984), 466-476.

J. SOKOLOWSKI [1], "Conical differentiability of projection on convex sets-an application to sensitivity analysis of Signorini variational inequaality," Technical Report, Institute of Mathematics of the University of Genova, 1981.

J. SOKOLOWSKI AND J.P. ZOLÉSIO [1], *Dérivée par rapport au domaine de la solution d'un problème unilatéral*, C.R. Acad. Sci. Paris, Sér. I **3301** (1985), 103-106.

[2], *Shape sensitivity analysis of unilateral problems*, SIAM J. on Math. Anal. **18** (1987), 1416-1437.

MICHEL C. DELFOUR
Centre de recherches mathématiques, Université de Montréal
CP 6128 – A, Montréal (Québec) H3C 3J7, CANADA

JEAN-PAUL ZOLÉSIO
Institut Non Linéaire de Nice Faculté des Sciences
Parc Valrose, F–06034 Nice –Cédex, FRANCE

S HUANG
# On the Stefan problem with nonlinear boundary conditions and its optimal control

In this talk we first consider the existence and uniqueness of weak solutions to an n–dimensional two–phase Stefan problem with nonlinear flux, and then we further consider the optimal control for this system. The study of these problems can be viewed as a model of a solidification process in which the flux is given by Stefan–Boltzmann's radiation law. Our results can be mainly compared with the work in [1]–[5].

Let $\Omega \subset R^n$ be a bounded domain with smooth boundary $\Gamma$, $T > 0, Q = \Omega \times (0,T)$, $\Sigma = \Gamma \times (0,T)$. The classical formulation of the problem we consider is as follows ([1]):

Problem (S)

Find a function $\theta(x,t)$ and a free boundary $S = \{(x,t) \in Q \,|\, \Phi(x,t) = 0\}$ such that

$$C(\theta)\,\frac{\partial\theta}{\partial t} - \operatorname{div}[k(\theta)\nabla\theta + \vec{b}_0(x,t,\theta)] = f_0(x,t,\theta) \;\; \text{in} \;\; Q \backslash S$$

$$K(\theta)\,\frac{\partial\theta}{\partial n} + \vec{b}_0(x,t,\theta)\cdot\vec{n} = g_0(x,t,\theta) \;\; \text{on} \;\; \Sigma$$

$$\theta(x,t) = 0 \;\; \text{on} \;\; S$$

$$\nu\,\frac{\partial\Phi}{\partial t} = [k(\theta)\nabla\theta + \vec{b}_0(x,t,\theta))\cdot\nabla\Phi]_-^+ \;\; \text{on} \;\; S$$

$$\theta(x,0) = \theta_0(x) \;\; \text{in} \;\; \Omega$$

where $\nu$ is a given positive constant, and $e$, $k$, $\vec{b}_0$, $f_0$, $g_0$, $\theta_0$ are given functions of their arguments.

For simplification, we assume that $\vec{b}_0(x,t,\theta) \equiv 0$. All the results in this talk can be extended to the case that $\vec{b}_0(x,t,\theta) \neq 0$.

Set

$$K(r) = \int_0^r k(\eta)d\eta, \;\; h(r) = \int_0^r \frac{c(K^{-1}(\eta))}{k(K^{-1}(\eta))}\,d\eta,$$

and

$$\gamma(r) = h(r) + \nu H(r), \tag{0.1}$$

where

$$H(r) = \begin{cases} 1, & r > 0, \\ [0,1], & r = 0, \\ 0, & r < 0, \end{cases}$$

and

$$y(x,t) = K(\theta(x,t)), \quad f(x,t,y) = f_0(x,t,K^{-1}(y)),$$

$$g(x,t,y) = g_0(x,t,K^{-1}(y)), \quad y_0(x) = K(\theta_0(x)).$$

Now as in [1], the weak formulation of Problem (S) is defined as follows:

Problem (P)

Find a pair $(y,w)$ such that

$$y \in L^2(0,T;H^1(\Omega)), \quad w \subset \gamma(y), \quad w \in L^2(0,T;L^2(\Omega)),$$

and

$$\int_Q \int \left[ w \frac{\partial\phi}{\partial t} - \nu y \cdot \nabla\phi + f(x,t,y)\phi \right] dxdt + \int_\Sigma g(x,t,y)\phi d\sigma + \int_\Omega w_0(x)\phi(x,0)dx = 0,$$

$$\forall \phi(x,t) \in H^1(Q), \quad \phi(x,T) = 0,$$

where $w_0 \subset \gamma(y_0)$, and $y_0(x)$ is given.

## §1 Existence and Uniqueness Theorem.

Assume that $h(r)$ is Lipschitz continuous, $h(0) = 0$, and there are constants $h_0$, $h_1$ such that $\hspace{3cm}$ (1.1)

$$0 < h_0 \leq h'(r) \leq h_1, \quad \text{a.e. for } r \in R^1,$$

and $\gamma(r)$ is defined by (0.1),

$f(x,t,r)$ and $g(x,t,r)$ satisfy Caratheodory conditions, i.e., are continuous in $r \in R^1$ for a.e. $(x,t) \in Q$ (or $\Sigma$), and measurable in $(x,t) \in Q$ (or $\Sigma$) for all $r \in R^1$ and in addition $f(x,t,0) \in L^p(Q)$, $g(x,t,0) \in L^p(\Sigma)$, $\hspace{1cm}$ (1.2)

$f(x,t,r)$ and $g(x,t,r)$ are Lipschitz continuous in $r \in R^1$, uniformly for a.e. $(x,t) \in Q$ (or $\Sigma$), $\hspace{4cm}$ (1.3)

$$w_0(x) \in L^p(\Omega), \tag{1.4}$$

where $p > 2$ is a constant.

Theorem 1.1 $\quad$ Under the assumptions (1.1)–(1.4), there exists one and only one solution $(y,w)$ of Problem (P), and it satisfies

$$y \in L^\infty(0,T; L^P(\Omega)), \quad \gamma_0 y \in L^P(\Sigma), \tag{1.5}$$

$$w \in L^\infty(0,T; L^P(\Omega)), \frac{\partial w}{\partial t} \in L^2(0,T; (H^1(\Omega))'), \tag{1.6}$$

where $\gamma_0$ is the trace operator on $\Gamma$. □

The existence result was discussed in [1]. Theorem 1.1 claims that the weak solution of Problem (S) in $L^\infty(0,T; L^P(\Omega))(p > 2)$ is unique.

Now we consider the case in which the assumption (1.3) can be relaxed.

Assume that

$f(x,t,r)$ and $g(x,t,r)$ satisfy Caratheodory conditions, $f(x,t,0) \in L^\infty(Q)$, $g(x,t,0)$ $\in L^2(\Sigma)$. And there is a constant $R_0 > 0$, such that

$$r \cdot g(x,t,r) \leq 0, \ \forall \ |r| \geq R_0, \ \text{a.e.} \ (x,t) \in \Sigma, \tag{1.2}*$$

$f(x,t,r)$ and $g(x,t,r)$ are Lipschitz continuous in $r \in [-R, R]$, uniformly for a.e. $(x,t) \in Q$ (or $\Sigma$), $\forall R > 0$, $\tag{1.3}*$

$$w_0(x) \in L^\infty(\Omega). \tag{1.4}*$$

<u>Theorem 1.2</u> Under the assumptions (1.1) and $(1.2)^* - (1.4)^*$, there exists one and only one solution $(y,w)$ of Problem (P), and it satisfies

$$y \in L^\infty(Q), \ \gamma_0 y \in L^\infty(\Sigma), \tag{1.7}$$

$$y \in L^\infty(Q), \frac{\partial w}{\partial t} \in L^2(0,T; (H^1(\Omega))'). \quad □ \tag{1.8}$$

This result was obtained in [4] where $f(x,t,r)$ and $g(x,t,r)$ were in addition assumed to be monotone with respect to $r$. No monotonicity hypothesis are required here.

The existence an uniqueness results stated above will be used to discuss the optimal control for this system.

## §2 Optimal Boundary Control

We are restricted to discussing the optimal boundary control for Problem (P) with assumptions (1.1)–(1.4). Similar results hold for Problem (P) as the following form:

$$g(x,t,y) = \bar{g}(x,t,y) + u(x,t), \tag{2.1}$$

where $u(x,t) \in L^P(\Sigma)$ is the control variable. This problem has been discussed in [3] where the authors consider the set of admissible controls $U_{ad} \subset H^1(0,T; L^2(\Gamma))$,

226

assumed to be closed, bounded and convex subset of $H^1(0,T; L^2(\Gamma))$, bounded as a subset of $L^\infty(\Sigma)$ ([3] p.415). In this talk, we will relax the restriction on $U_{ad}$. Let

$U_{ad}$ be a closed, bounded and convex subset of $L^p(\Sigma)$. $\qquad\qquad$ (2.2)

By $(y(u), w(u))$ we shall denote the response of Problem (P), corresponding to the control $u \in U_{ad}$.

<u>Lemma 2.1</u>  Let $u_n \longrightarrow u$ weakly in $L^p(\Sigma)$. Then

$$y(u_n) \longrightarrow y(u) \text{ strongly in } L^2(Q),$$

$$w(u_n) \longrightarrow w(u) \text{ weakly}^* \text{ in } L^\infty(0,T; L^p(\Omega)),$$
$$\text{strongly in } L^2(0,T; (H^1(\Omega))'). \qquad \square$$

For simplification, we shall consider only one type of optimal control problem formulated for the Stefan process:

$(P_Q)$  minimize $J(u) \equiv I[y(u)]$ over $U_{ad}$.

The functional $I$ is assumed to be lower semicontinuous over $L^2(Q)$.

We can claim the existence of optimal controls.

<u>Theorem 2.2</u>  There exists an optimal pair $(u, y(u))$ for Problem $(P_Q)$ $\qquad \square$.

Now we discuss the approximations of Problem $(P_Q)$. Let $\gamma_\epsilon$, $f_\epsilon$ and $g_\epsilon$ be $\varphi$ suitable smooth approximations to the data of Problem (P).

<u>Problem $(P_\epsilon)$</u>  Find $y_\epsilon \in L^2(0,T; H^1(\Omega))$, such that

$$\int_Q \int \left[ \gamma_\epsilon(y_\epsilon) \frac{\partial \phi}{\partial t} - \nabla y_\epsilon \cdot \nabla \phi + f_\epsilon(x,t,y_\epsilon)\phi \right] dxdt$$

$$+ \int_\Sigma [g_\epsilon(x,t,y_\epsilon) + u(x,t)]\phi d\sigma + \int_\Omega \gamma_\epsilon(y_0)\phi(x,0)dx = 0,$$

$$\forall \phi \in H^1(Q), \phi(x,T) = 0,$$

where $y_0 = \gamma^{-1}(w_0)$.

Now we consider the following sequences of optimal control for Problem $(P_\epsilon)$:

$(P_Q^\epsilon)$  minimize $J_\epsilon(u) \equiv I[y_\epsilon(u)]$ over $U_{ad}$.

One can claim that there exists one and only one solution of Problem $(P_\epsilon)$, and $(P_Q^\epsilon)$ has an optimal pair $(u_\epsilon, y_\epsilon(u_\epsilon))$.

<u>Theorem 2.3</u> Let the functional I be continuous over $L^2(Q)$. Then on a subsequence we have the convergencies:

$$u_\epsilon \rightharpoonup u \text{ weakly in } L^p(\Sigma),$$

$$y_\epsilon(u_\epsilon) \rightarrow y(u) \text{ strongly in } L^2(Q),$$

where $(u, y(u))$ is an optimal pair for Problem $(P_Q)$. [ ]

## REFERENCES

[1] J.R. Cannon and E. DiBenedetto, *On the existence of weak–solutions to an n–dimensional Stefan problem with nonlinear boundary conditions*, SIAM J. Math. Anal., Vol. 11, No. 4(1980), pp 632–645.

[2] M. Neizgodka and I. Pawlow, *A generalized Stefan problem in several space variables*, Appl. Math. Optim., 9(1983), pp 193–224.

[3] M. Niezgodka and I. Pawlow, *Optimal control for parabolic systems with free boundaries–existence of optimal controls, approximation results*, Proc. 9th IFIP conference Optim Techniqes. Springer–Verlag, New York–Heidelberg–Berlin. Lecture Notes Control and Information Sci 22: 412–421, (1980).

[4] M. Niezgodka and I. Pawlow and A. Visintin, *Remarks on the paper by A. Visintin "Sur le probleme de Stefan avec flux non lineaire"*, Bollettino U.M.I., Analisi Funzionale e Applicazioni, Serie V. Vol. XVIII–C.N. 1–1981, pp 87–88.

[5] A. Visintin, *Sur le probleme de Stefan avec flux non lineaire*, Bollettino U.M.I., Analisi Funzionale e Applicazioni Serie V. Vol XVIII–C.N. 1–1981, pp 63–86.

**Shaoyun Huang**
Department of Mathematics
Peking University

M D P M MARQUES

# Minimization of functionals on classes of Lipschitz domains and *BV* functions

We analyse the extension of recent minimization results for functionals defined on Sobolev spaces with changing Lipschitz domains (like in [MM]) to the setting of $BV$ functions. To be precise, following Ambrosio[A], we introduce a class, denoted by $SBV$, of special real functions of bounded variation $u$, whose distributional derivatives $Du$ are sums of $(n-1)$-dimensional measures with absolutely continuous measures with respect to the Lebesgue measure in $\mathbb{R}^n$ (see also [DGA] for a general definition for vector-valued functions). The domains of "integration" are allowed to vary in a class of Lipschitz domains with uniform cone property, denoted by $\mathcal{C}_\theta^{0,1}(D)$. This class was introduced by Chenais[C] to study some domain identification problems.

Consider the functional

$$(1) \qquad F(\Omega, u) := \int_\Omega [\tilde{f}(x, u, \nabla u) + |u - w|^\alpha] \, dx + \int_{S_u} |u^+ - u^-|^\beta \, d\mathcal{H}_{n-1} \,,$$

where $\alpha > 0$, $0 \leq \beta < 1$, $w$ is a given Borel function and the normal convex integrand $\tilde{f}$ verifies some standard requirements. Here $\nabla u$ denotes the approximate differential of the $SBV$ function $u : \Omega \to \mathbb{R}$; $u^+$ and $u^-$ are the approximate upper and lower limits of $u$, $S_u$ is the set of points where $u^- < u^+$ and $\mathcal{H}_{n-1}$ is the $(n-1)$-dimensional Hausdorff measure. As pointed out by Ambrosio, $SBV$ is a natural domain because: 1) the functional depends only on the approximate differential and on the set of "jumps" $S_u$ ; and 2) in the physical problems which it models, e. g. in liquid crystals theory or in pattern recognition and computer vision, the solutions are expected to be reasonably regular outside an unkown but at most $(n-1)$-dimensional set of "cracks". It is worth noting that in the functional (1) is not involved a convex function of the measure $Du$ (as defined e. g. in [H][KT][T][TD]) for which l.s.c. theorems are also available. Instead we deal here with a convex function of the $\nabla u \, dx$ and a concave function of the jump part (when $0 < \beta < 1$).

When $\Omega$ is fixed, the existence of minimizer for the functional (1) is deduced in [A] (without a detailed proof) from a compactness and lower semicontinuity theorem established therein and from lsc results of Ioffe. For the simpler functional

(2)
$$G(\Omega, u) = \int_{\Omega} f(x, u, \nabla u) \, dx + l(\Omega) \quad ,$$

where $l: \mathbb{C}_\theta^{0,1}(D) \to \mathbb{R}$ is a lsc function, it was proved in [MM] that there exist minimizers $(\Omega, u)$ with $\Omega \varepsilon \, \mathbb{C}_\theta^{0,1}(D)$ and $u \varepsilon \, W^{1,1}(\Omega)$ satisfying boundary and obstacle conditions: $u = v$ on $\partial\Omega$ and $u \geq \phi$ in $\Omega$.

Similarly, we shall consider a class of functionals which includes (1) as a particular case:

(3)
$$F(\Omega, u) := \int_{\Omega} f(x, u, \nabla u) \, dx + \int_{S_u} \psi \, (u^+ - u^-) \, d\mathcal{H}_{n-1} \quad ,$$

under the assumptions and definitions listed below.

• $D \subset \mathbb{R}^n$ is a fixed *Lipschitz domain* of class $\mathbb{C}^{0,1}$, that is a domain (nonempty bounded connected open subset) which locally is the hypograph of Lipschitz continuous functions in adequate systems of orthogonal coordinates.

• $\Omega$ belongs to $\mathbb{C}_\theta^{0,1}(D)$ the class of Lipschitz subdomains $\Omega$ of $D$ verifying the uniform cone property with a fixed parameter $\theta \varepsilon \, ]0, \pi/2[$. This means that for every $x \varepsilon \partial\Omega$ there is a unit vector $\xi_x$ such that $\Omega$ contains every set of the form $y + C_x$ , where $y$ belongs to the $\theta$-neighbourhood of $x$ and $C_x$ is the open (round) cone with axis $\xi_x$ , angle $\theta$ and height $2\theta$ :
$C_x = \{ z \varepsilon \mathbb{R}^n : ||z|| < 2\theta \,, \, z . \xi_x > ||z|| \cos \theta \, \}$.

• $BV(\Omega)$ is the class of real functions $u : \Omega \to \mathbb{R}$ with bounded variation, having Radon $n$-dimensional measures as distributional derivatives, denoted $Du$. They are (approximately) differentiable almost everywhere and $\nabla u$ coincides a.e. with the Radon-Nikodim derivative $Du/dx$, where it exists.

• $SBV(\Omega)$ is the class of special functions with bounded variation, for which $Du$ is the sum of $\nabla u \, dx$ with a measure having a density with respect to the $(n-1)$-dimensional Hausdorff measure on the set of jumps $S_u$ . Hence $Du = \nabla u \, dx + Ju$, where $Ju(B) := Du(B \cap S_u)$ for every Borel set $B$.

• $GBV(\Omega)$ , respectively $GSBV(\Omega)$, is the class of functions $u$ such that every truncated function $N \wedge u \vee (-N)$ with $N \varepsilon \mathbb{N}$ belongs to $BV(\Omega)$, respectively to $SBV(\Omega)$.

**Hypothesis H1** - The function $f(x, u, z) : D \times \mathbb{R} \times \mathbb{R}^n \to [0, +\infty]$ is a *normal convex integrand*, i.e. it is a Borel function, lower semicontinuous in $(u, z)$ and convex in $z$; moreover, $f$ verifies the following coerciveness condition:

(4)
$$f(x, u, z) \geq \phi(||z||) + g(x, u) \quad ,$$

where $\phi:[0,+\infty[ \to [0,+\infty[$ is a convex nondecreasing function with superlinear growth ($\phi(t)/t \to +\infty$ as $t \to +\infty$) and $g:D \times \mathbb{R} \to [0,+\infty]$ is a Borel function, lower semicontinuous in $u$ ( a normal integrand), coercive in $u$: $g(x,u) \to +\infty$ as $|u| \to +\infty$.

**Hypothesis H2** - The function $\psi:]0,+\infty] \to [0,+\infty]$ is a concave nondecreasing function such that

$$\psi(0) = 0 \ , \ \lim_{t \downarrow 0} \frac{\psi(t)}{t} = +\infty \quad \text{and} \quad \psi(+\infty) = \lim_{t \to +\infty} \psi(t) \ .$$

Remark that to get (1) we can take $f(x,u,z) = \tilde{f}(x,u,z) + |u-w|^\alpha$, $g(x,u) = |u-w|^\alpha$ and $\psi(t) = t^\beta$, with $0 < \beta < 1$. If $\tilde{f}(x,u,z) = ||z||^\gamma$ with $\gamma > 1$ we obtain the particular case [A] (0.1):

$$(5) \qquad F(\Omega,u) = \int_\Omega ||\nabla u||^\gamma \, dx + \int_\Omega |u-w|^\alpha \, dx + \int_{S_u} |u^+ - u^-| \, d\mathcal{H}_{n-1} \ ,$$

but the choice of $\gamma = 2$ and $\beta = 0$ giving as in [A] (2.8)

$$(6) \qquad F(\Omega,u) = \int_\Omega ||\nabla u||^2 \, dx + \mathcal{H}_{n-1}(S_u) + \int_\Omega |u-w|^\alpha \, dx \ ,$$

is not allowed here; the latter occurs in computer vision theory.

This type of functional apparently does not behave nicely with respect to minimization subject to boundary conditions. A kind of relaxation seems to be in order (as often is the case in the works of Temam, Strang, Demengel and Hadhri, e.g. [TS] [D] [H2]). This *relaxation* is done here by adding to $F$ the term

$$(7) \qquad \int_{\partial\Omega} \psi(|\gamma_\Omega u - h|) \, d\mathcal{H}_{n-1} \ ,$$

which measures in some sense the distance between a preassigned boundary value function $h$ and the (inner) trace of $u$ on the boundary of $\Omega$, $\gamma_\Omega u \varepsilon L^1(\partial\Omega)$; this exists even on Lipschitz boundaries, cf. [F] [G]. Moreover we restrict the search for minimizers by demanding that they take values in some fixed compact interval, say $[0,M]$. A first result in this direction of research is the following:

**THEOREM** - *Let $D$ be a Lipschitz domain and fix $\theta \varepsilon ]0,\pi/2[$. Assume that $f$ and $\psi$ satisfy hypotheses H1 and H2. Let $h \varepsilon W^{1,\infty}(D)$ be given with*

$$(8) \qquad 0 \le h(x) \le M \ (x \varepsilon D) \text{ and } \int_D f(x,h,\nabla h) \, dx < +\infty \ .$$

*Consider the functional*

$$(9) \qquad \tilde{F}(\Omega,u) := \int_\Omega f(x,u,\nabla u) \, dx + \int_{S_u} \psi(u^+ - u^-) \, d\mathcal{H}_{n-1} + \int_{\partial\Omega} \psi(|\gamma_\Omega u - h|) \, d\mathcal{H}_{n-1}.$$

231

*Then there is a minimizer $(\Omega_0, u_0)$ for $\tilde{F}$ in the class*

(10) $$\mathcal{U} = \{ (\Omega, u) \mid \Omega \, \varepsilon \, \mathcal{C}_{\theta}^{0,1}(D), \ u \, \varepsilon \, SBV(\Omega) \} \ ,$$

*subject to the restriction*

(11) $$0 \leq u(x) \leq M \quad a. \ e. \ in \ \Omega \ .$$

**Proof.** One begins by extending $f$ to a larger set $D' = D + B(0, r)$ with sufficiently small $r$ in such a way that $f \colon D' \times \mathbb{R} \times \mathbb{R}^n \to [0, +\infty]$ still verifies the assumption H1 (this is done by partition of unity and local reflection with respect to the graphs describing the boundary of $\Omega$). The same extension operation is done with $g$ and $h$ so that (8) remains true in the enlarged set $D'$.

Take a minimizing sequence $(\Omega_m, u_m)$, $m \geq 1$ ; that is:

(12) $$\tilde{F}(\Omega_m, u_m) \to b := \inf \{ \, \tilde{F}(\Omega, u) \mid (\Omega, u) \, \varepsilon \, \mathcal{U}, \ 0 \leq \overset{.}{u} \leq M \ a. \ e. \, \} \ ,$$

where it is clear that $0 \leq b < +\infty$ and it may be assumed that

(13) $$\tilde{F}(\Omega_m, u_m) \leq K_1 < +\infty \ .$$

By a compactness theorem of Chenais (see [C] or [R] theorems 3:4.4 and 3:4.6) we may extract a subsequence still denoted by $(\Omega_m)$ which converges to a set $\Omega_0 \, \varepsilon \, \mathcal{C}_{\theta}^{0,1}(D)$ in the sense : $d_{\mathcal{L}}(\Omega_m, \Omega_0) := \int_D | \, \chi_{\Omega_m} - \chi_{\Omega_0} | \, dx \to 0$ (Lebesgue convergence). It is also known that in such a situation $(\overline{\Omega}_m)$ converges to $\overline{\Omega}_0$ in the sense of Hausdorff distance and $(\partial \Omega_m)$ converges to $\partial \Omega_0$ in Hausdorff distance and locally in the sense of Lipschitz graphs. In particular for some $K_2 < +\infty$:

(14) $$\mathcal{H}_{n-1}(\Omega_m) \leq K_2 \qquad (m \geq 1) \ .$$

Extend every $u_m$ to a function $v_m$ defined on $D'$ by putting $v_m(x) := h(x)$ if $x \, \varepsilon \, D' \backslash \Omega_m$ . Then $v_m \, \varepsilon \, SBV(D')$, $S_{v_m} \subset S_{u_m} \cup \partial \Omega_m$ and in $\partial \Omega_m$ we have $v_m{}^+ - v_m{}^- = | \gamma_{\Omega_m} u_m - h |$ . Defining for $v \, \varepsilon \, SBV(D')$:

(15) $$H(v) := \int_{D'} [\phi(\|\nabla v\|) + g(x, v)] \, dx + \int_{S_v} \psi(v^+ - v^-) \, d\mathcal{H}_{n-1} \ ,$$

we observe that, thanks to our hypothesis

$$\begin{aligned}
H(v_m) &\leq \int_{D'} f(x, v_m, \nabla v_m) \, dx + \int_{S_{u_m}} \psi(u_m{}^+ - u_m{}^-) \, d\mathcal{H}_{n-1} \\
&\quad + \int_{S_{v_m} \cap \partial \Omega_m} \psi(| \gamma_{\Omega_m} u_m - h |) \, d\mathcal{H}_{n-1} \\
&\leq \tilde{F}(\Omega_m, u_m) + \int_{D' \backslash \Omega_m} f(x, h, \nabla h) \, dx + \psi(M) \, \mathcal{H}_{n-1}(\partial \Omega_m).
\end{aligned}$$

So, taking $K := K_1 + \int_{D'} f(x, h, \nabla h) \, dx + \psi(M) K_2$, we have

(16) $\qquad\qquad \forall m: \quad H(v_m) \leq K < +\infty$ .

Since $\phi$, $g$ and $\psi$ satisfy the assumptions of the compactness theorem of Ambrosio ([A] theorem 2.1) we deduce from (16) that there exists a subsequence of $(v_m)$ — with unchanged notation — such that $(v_m)$ converges a. e. in $D'$ to a function $v_0 \, \varepsilon \, GSBV(D')$, the approximate differentials $\nabla v_m$ weakly converge to $\nabla v_0$ in $L^1(D', \mathbb{R}^n)$ and

(17) $\qquad \int_{S_{v_0}} \psi(v_0{}^+ - v_0{}^-) \, d\mathcal{H}_{n-1} \leq \liminf_m \int_{S_{v_m}} \psi(v_m{}^+ - v_m{}^-) \, d\mathcal{H}_{n-1}$ .

It is clear that $0 \leq v_m \leq M$ implies $0 \leq v_0 \leq M$ so that in fact $v_0 \, \varepsilon \, SBV(D')$ and $v_m \to v_0$ in $L^1(D)$ strong.

Define $u_0 := v_0|_{\Omega_0}$. Of course, $u_0 \, \varepsilon \, SBV(\Omega_0)$ and $0 \leq u_0 \leq M$ a. e. . Remark that

$$\int_{\Omega_m} f(x, u_m, \nabla u_m) \, dx = \int_{\Omega_0} f(x, v_m, \nabla v_m) \, dx - \int_{\Omega_0 \setminus \Omega_m} f(x, v_m, \nabla v_m) \, dx$$
$$+ \int_{\Omega_m \setminus \Omega_0} f(x, u_m, \nabla u_m) \, dx \, ;$$

that $\int_{\Omega_0 \setminus \Omega_m} f(x, v_m, \nabla v_m) \, dx = \int_{\Omega_0 \setminus \Omega_m} f(x, h, \nabla h) \, dx \quad \to \quad 0$, because of (8) and

$\text{meas}(\Omega_0 \setminus \Omega_m) \to 0$; and that $f \geq 0$ implies $\liminf_m \int_{\Omega_m \setminus \Omega_0} f(x, u_m, \nabla u_m) \, dx \geq 0$. Thus:

(18) $\quad \liminf_m \int_{\Omega_m} f(x, u_m, \nabla u_m) \, dx \geq \liminf_m \int_{\Omega_0} f(x, v_m, \nabla v_m) \, dx \geq \int_{\Omega_0} f(x, u_0, \nabla u_0) \, dx,$

applying a lower semicontinuity theorem of Olech (see [I]) as in [MM], because $(v_m, \nabla v_m)$ strong-weak converges and $f$ is a normal convex integrand that verifies by hypothesis, for every $r > 0$ :

$$f(x, u, z) \geq r||z|| - \eta(x, r) - a|u| \, ,$$

where $a = 0$ and $\eta(x, r) := |\min \{ \phi(t) - rt \mid t \geq 0 \}|$ is a normal integrand, summable on $D$ for all $r \geq 0$.

Concerning the jump terms it suffices to remark that in (17), for $m \geq 0$ :

$$\int_{S_{v_m}} \psi(v_m{}^+ - v_m{}^-) \, d\mathcal{H}_{n-1} = \int_{S_{u_m}} \psi(u_m{}^+ - u_m{}^-) \, d\mathcal{H}_{n-1} + \int_{\partial \Omega_m} \psi(|\gamma_{\Omega_m} u_m - h|) \, d\mathcal{H}_{n-1} \, .$$

because in $\partial \Omega_m \setminus S_{v_m}$ we have $\psi(v_m{}^+ - v_m{}^-) = \psi(0) = 0$. So from (17) and (18) it results that:

$$\tilde{F}(\Omega_0, u_0) \leq \liminf_m \int_{\Omega_m} f(x, u_m, \nabla u_m)\, dx + \liminf_m \left[ \int_{S_{u_m}} \psi(u_m{}^+ - u_m{}^-)\, d\mathcal{H}_{n-1} \right.$$

$$\left. + \int_{\partial\Omega_m} \psi(|\gamma_{\Omega_m} u_m - h|)\, d\mathcal{H}_{n-1} \right]$$

$$\leq \liminf_m \tilde{F}(\Omega_m, u_m) = b\,.$$

Hence $(\Omega_0, u_0)$ is the sought minimizer. $\qquad\qquad\square$

## REFERENCES

[A] AMBROSIO, L. - Compactness for a special class of functions of bounded variation, preprint Scuola Normale Superiore, Pisa.

[C] CHENAIS, D. - On the existence of a solution in a domain identification problem, J. Math. Anal. Appl., 52 (1975), 189-219.

[DGA] DE GIORGI, E. and AMBROSIO, L.- Un nuovo tipo di funzionale del Calcolo delle Variazioni, Atti Accad. Naz. Lincei Rend. Cl. Sci. Fis. Mat. Natur.

[D] DEMENGEL, F. - Some compactness theorems for spaces of functions with bounded derivatives and applications to limit analysis problems in plasticity, preprint 87 T 9, Université de Paris-Sud.

[F] FERRO, F.- $BV$ spaces on manifolds and functions whose traces are measures, Boll. U. M. I. (6)4-B (1985) 211-228.

[G] GIUSTI, E.- *Minimal surfaces and functions of bounded variation*, Australian National University, Canberra, 1977.

[H] HADHRI, T.- Fonction convexe de mesure, C. R. Acad. Sci. Paris, t. 301, Sér. 1, n° 13 (1985).

[H2] HADHRI, T.- Étude dans $HB \times BD$ d'un modèle de plaques élastoplastiques comportant une non-linéarité géométrique, R. A. I. R. O. M2AN, 19 (1985) 235-283.

[I] IOFFE, A. D. - On lower semicontinuity of integral functionals I , SIAM J. Control Optimization, 15 (1977), 521-538.

[KT] KOHN, R. and TEMAM, R.- Dual spaces of stresses and strains with applications to Hencky plasticity, Appl. Math. and Optimization, 10 (1983) 1-35.

[MM] MONTEIRO MARQUES, M.- Minimization of integral functionals depending on Lipschitz domains, Numer. Funct. Anal. and Optimiz., 10 (1989) 991-1002.

[R] RODRIGUES, J.-F. - *Obstacle Problems in Mathematical Physics* , North-Holland Math. Studies, Amsterdam, 1987.

[T] TEMAM, R.- *Problèmes variationnels en plasticité*, Gauthiers-Villars, Paris, 1983.

[TD] TEMAM, R. and DEMENGEL, F. - Convex function of a measure, Indiana Journal of Mathematics, 33 (1984) 673-709.

[TS] TEMAM, R. and STRANG, G.- Duality and relaxation in the variational problems of plasticity, Journal de Mécanique, 19 (1980) 493-527.

## ACKNOWLEDGMENT
This research was partly supported by contract 87046 JNICT/INIC.

Faculdade de Ciências de Lisboa and Centro de Matemática e Aplicações Fundamentais /I.N.I.C.

M RAO AND J SOKOLOWSKI
# Stability of solutions to unilateral problems

## 1. INTRODUCTION

We provide results on differential stability of metric projection in Sobolev space $H_0^{2m}(\Omega)$ onto convex set

$$K = \{f \in H_0^{2m}(\Omega) \mid f(x) \geq \psi(x), x \in \Omega\} \tag{1.1}$$

where $\Omega \subset R^d$ is an open, bounded domain. We refer the reader to [2], [5] for the related results in the Sobolev space $H_0^1(\Omega)$. Some applications of the differential stability results for variational inequalities are presented in [3], [6] - [14]. We recall some properties of Sobolev spaces and the notion of capacity [16]. The Sobolev space $H_0^{2m}(\Omega)$ is the closure of $C_0^\infty(\Omega)$ with norm

$$\|\varphi\|_{H_0^{2m}(\Omega)}^2 = \int_\Omega |\Delta^m \varphi|^2 dx$$

Functions in $H_0^{2m}(\Omega)$ are defined quasi everywhere and are quasi continuous. These notions are made precise below.

The $C_{2m}$–capacity of a compact set $F$ is defined as

$$C_{2m}(F) = \inf\{ \int |\Delta^m \varphi|^2 dx \; : \; \varphi \geq 1 \text{ on } F, \; 0 \leq \varphi \in C_0^\infty(R^d) \}$$

The capacity of a Borel set is then defined as the supremum of capacities of its compact subsets. A statement holds $C_{2m}$–q.e., if it holds except for a set of $C_{2m}$–capacity zero. With this definition we have the following results:

1. Let $\varphi \in H_0^{2m}(\Omega)$, and $\{\varphi_n\} \subset C_0^\infty(\Omega)$ converge to $\varphi$ in $H_0^{2m}(\Omega)$. Then a subsequence of $\{\varphi_n\}$ converge $C_{2m}-$ q.e. and this is a representative of $\varphi$.
2. Let $\varphi \in H_0^{2m}(\Omega)$. Then $\varphi$ has a quasicontinuous representative: There is a representative $\overline{\varphi}$ such that given $\varepsilon > 0$, there is an open set $U(\varepsilon)$ of $C_{2m}$–capacity less than $\varepsilon$ such that the restriction of $\overline{\varphi}$ to the complement of $U(\varepsilon)$ is continuous.
3. Any two quasi continuous representatives of $\varphi \in H_0^{2m}(\Omega)$ agree $C_{2m}$–q.e.
4. Every set of positive Lebesque measure has positive $C_{2m}$–capacity.

We refer the reader to [16] for further results on capacity in Sobolev spaces. Standard notation throughtout the paper [1],[16].

## 2. TANGENT CONE

We shall consider the metric projection onto the following convex set

$$K = \{f \in H_0^{2m}(\Omega) \mid f(x) \geq \psi(x), x \in \Omega\} \tag{2.1}$$

with respect to the scalar product

$$(y, z) = \int_\Omega \Delta^m y(x) \Delta^m z(x) dx \tag{2.2}$$

We assume that $\psi \in H^{2m}(\Omega)$, $\psi(x) < 0$ on $\partial\Omega$, therefore set (2.1) is nonempty . The metric projection $z = P_K y$ , $y \in H_0^{2m}(\Omega)$, is given by the unique solution of the following variational inequality

$$z \in K \; : \; \int_\Omega \Delta^m z(x) \Delta^m(\varphi - z)(x) dx \geq \int_\Omega \Delta^m y(x) \Delta^m(\varphi - z)(x) dx \tag{2.3}$$

$$\forall \varphi \in K$$

We denote

$$C_K(z) = \{\varphi \in H_0^{2m}(\Omega) \mid \exists t > 0 \text{ such that } z + t\varphi \in K\} \tag{2.4}$$

We derive the form of tangent cone $T_K(z) = \text{cl} C_K(z)$ for any element $z$ in convex set (2.1).

## THEOREM 1

For any element $z \in K$, tangent cone $T_K(z)$ takes the form

$$T_K(z) = \{\varphi \in H_0^{2m}(\Omega) \mid \varphi(x) \geq 0 \, ; \, C_{2m} - \text{q.e. on } \Xi\} \tag{2.5}$$

where $\Xi = \{x \in \Omega \mid z(x) = \psi(x)\} \subset \Omega$.

## PROOF OF THEOREM 1

Note that $C_K(z)$ and hence also $T_K(z)$ is a convex cone containing all non-negative elements of $H_0^{2m}(\Omega)$. Let an element $V \in H_0^{2m}(\Omega)$ be given and suppose that $V \geq 0$ $C_{2m}$−q.e. on $\Xi$. There exists a unique element $\phi_0 \in T_K(z)$ such that

$$\|V - \phi_0\|_{H_0^{2m}(\Omega)}^2 = \inf \{\|V - \phi\|_{H_0^{2m}(\Omega)}^2 \mid \phi \in C_K(z)\} \tag{2.6}$$

It is easy to see that for any $H_0^{2m}(\Omega) \ni \phi \geq 0$ , $t \geq 0$ , $\phi_0 + t\phi \in T_K(z)$. Using (2.6) and standard arguments it follows

$$(V - \phi_0, \phi)_{H_0^{2m}(\Omega)} \leq 0 , \, 0 \leq \phi \in H_0^{2m}(\Omega) \tag{2.7}$$

hence there exists a non-negative Radon measure $\mu$ on $\Omega$ such that

$$(V - \phi_0, \phi)_{H_0^{2m}(\Omega)} = -\int \phi d\mu \, , \ \phi \in C_0^\infty(\Omega) \tag{2.8}$$

This implies in particular that for $\phi \geq 0$

$$\int \phi d\mu = -(V - \phi_0, \phi)_{H_0^{2m}(\Omega)} \leq \|V - \phi_0\|_{H_0^{2m}(\Omega)} \|\phi\|_{H_0^{2m}(\Omega)}$$

So by definition of $C_{2m}$−capacity we see $\mu$ cannot charge sets of zero $C_{2m}$−capacity. Since the measure may be large near the boundary it is not clear that (2.8) holds for all $\phi \in H_0^{2m}(\Omega)$. We can circumvent this difficulty by repeated use of a result of L. I. Hedberg: THEOREM 3.1 in [4]. First we show that (2.8) holds for any bounded $\phi \in H_0^{2m}(\Omega)$ which has compact support. Indeed for suitable mollifiers $\varrho_n$ , $\phi \star \varrho_n \in C_0^\infty(\Omega)$, have compact support, and tend boundedly pointwise $C_{2m}$-q.e. and in $H_0^{2m}(\Omega)$ to $\phi$. Since $\mu$ is Radon measure we may appeal to Lebesque dominated convergence to finish the claim . In the general case if $0 \leq \phi \in H_0^{2m}(\Omega)$ by the above theorem of Hedberg , we can select $0 \leq w_k \leq 1$, $k = 1, 2, ...$ such that $w_k \phi$ has compact support and is in $L^\infty$ approximating $\phi$ in $H_0^{2m}(\Omega)$. In particular $w_k \phi$ converges to $\phi$ $C_{2m}$-q.e. By (2.8) we have

$$\int w_k \phi d\mu = -(V - \phi_0, w_k \phi)_{H_0^{2m}(\Omega)}$$

is bounded, so by Fatou Lemma $\phi \in L^1(\mu)$. On the other hand $w_k \phi \leq \phi$ so dominated convergence applies

$$-\int \phi d\mu = (V - \phi_0, \phi)_{H_0^{2m}(\Omega)} \, , \ 0 \leq \phi \in H_0^{2m}(\Omega) \tag{2.9}$$

Now let $\phi \in C_0^\infty(\Omega)$ , $0 \leq \phi \leq 1$, then $\phi(z - \psi) \in H_0^{2m}(\Omega)$. We show that

$$\phi_0 + t\phi(z - \psi) \in T_K(z), \quad -1 \leq t \leq 1$$

It is sufficient to show that for any $\varphi \in C_K(z)$, $\varphi + t\phi(z - \psi) \in C_K(z)$. Now $\varepsilon\varphi + z - \psi \geq$ in $\Omega$ for some $\varepsilon > 0$, hence for $s > 0$, $\frac{s}{1-s} < \varepsilon$ we have

$$s[\varphi + t\phi(z - \psi)] + z - \psi \geq 0, \text{ in } \Omega$$

since $(1 + st\phi)(z - \psi) \geq (1 - s)(z - \psi)$. Using this in (2.6) with $\phi$ replaced by $\phi_0 + t\phi(z - \psi)$ we obtain

$$(V - \phi_0, \phi(z - \psi))_{H_0^{2m}(\Omega)} = 0$$

which, because $\phi(z - \psi)$ has compact support and belongs to $H_0^{2m}(\Omega)$ means

$$\int \phi(z - \psi) d\mu = 0$$

237

hence

$$\mu(x \ : \ z > \psi) = 0$$

i.e. $\mu$ is concentrated on $\Xi$. Our next step is to show that $\phi_0 = 0$ $\mu$-a.e. To this end using the fact that $T_K(z)$ is a cone and taking $t\phi_0$ for $\phi$ in (2.6) we get

$$(V - \phi_0, \phi_0)_{H_0^{2m}(\Omega)} = 0 \tag{2.10}$$

Now we use Hedberg's result once more. Choose $w_k$, $0 \le w_k \le 1$ such that $w_k\phi_0$ has compact support and converges to $\phi_0$ in $H_0^{2m}(\Omega)$. Since $\phi_0 \ge 0$ on $\Xi$ and $\mu$ is concentrated on $\Xi$, $w_k\phi_0 \le \phi_0$ $\mu$-a.e. So using the same argument as above we get

$$0 = (V - \phi_0, \phi_0)_{H_0^{2m}(\Omega)} = - \int \phi_0 d\mu$$

i.e. that $\phi_0 = 0$ $\mu$-a.e.

Finally since $\phi_0 = 0$ $\mu$-a.e and $V \ge 0$ $C_{2m}$-q.e. on $\Xi$ we can repeat the above argument to get

$$(V - \phi_0, V - \phi_0)_{H_0^{2m}(\Omega)} = - \int (V - \phi_0) d\mu = - \int V d\mu$$

But the right hand side is $\le 0$ because $V \ge 0$, thus $V = \phi_0$.

## 3. DIFFERENTIABILITY OF METRIC PROJECTION

We derive a result on the differentiability of metric projection $P_K$ in the Hilbert space $H = H_0^{2m}(\Omega)$ onto convex closed set $K \subset H$ of the form (2.1). We use the following notation. For any given element $u \in K$ we denote

$$C_K(u) = \{\phi \in H \mid \exists t > 0 \text{ such that } u + t\phi \ge \psi\} \tag{3.1}$$

The tangent cone $T_K(u)$ to $K$ at $u$ is the closure of set (3.1)

$$T_K(u) = \text{cl}(C_K(u)) \tag{3.2}$$

here cl stands for the closure. For a given element $g \in H_0^{2m}(\Omega)$, such that $f = P_K(g)$ let us define the following convex cone in the space $H_0^{2m}(\Omega)$

$$S = T_K(f) \cap [g - P(g)]^\perp = T_K(f) \cap [f - g]^\perp \tag{3.3}$$

We recall the following result on directional differentiability of metric projection onto $K$ [2],[5].

## LEMMA 1

Assume that for given $g \in H_0^{2m}(\Omega)$ the following condition is satisfied

$$S = \operatorname{cl}(C_K(f) \cap [f - g]^\perp) \tag{3.4}$$

then for all $h \in H_0^{2m}(\Omega)$ and for $t > 0$, t small enough

$$P_K(g + th) = P_K g + t P_S h + o(t)$$

## DEFINITION

Compact $F$ is admissible if for any element $\varphi \in H_0^{2m}(\Omega)$, $\varphi = 0$ on $F$ it follows $\varphi \in H_0^{2m}(\Omega \setminus F)$

Denote by $\nu \geq 0$ Radon measure defined as follows

$$\int \varphi d\nu = (f - g, \varphi)_{H_0^{2m}(\Omega)}, \quad \forall \varphi \in C_0^\infty(\Omega)$$

## LEMMA 2

If $F = \operatorname{spt}\nu$ is admissible condition (3.5) is satisfied.

## REFERENCES

[1] ADAMS, R.A. *Sobolev Spaces.* Academic Press, New York (1975)

[2] HARAUX, A. How to differentiate the projection on a convex set in Hilbert space.Some applications to variational inequalities. *J. Math. Soc. Japan* 29, (1977), 615-631

[3] HAUG, E.J. AND CEA, J. (EDS.) *Optimization of Distributed Parameter Structures.* Sijthoff and Noordhoff, Alpen aan den Rijn, The Netherlands, (1981)

[4] HEDBERG, L.I. Spectral Synthesis in Sobolev Spaces, and Uniqueness of Solutions of Dirichlet Problem. *Acta Math.* 147, (1981), 237-264

[5] MIGNOT, F. Controle dans les inequations variationelles elliptiques. *J. Funct. Anal.* 22, (1976), 25-39 .

[6] RAO, M. AND SOKOŁOWSKI, J. Sensitivity of unilateral problems in $H_0^2(\Omega)$ and applications. *to appear.*

[7] RAO, M. AND SOKOŁOWSKI, J. Shape sensitivity analysis of state constrained optimal control problems for distributed parameter systems. *Lecture Notes in Control and Information Sciences* Vol.114, Springer Verlag, (1989), 236-245

[8] RAO, M. AND SOKOŁOWSKI, J. Differential Stability of Solutions to Parametric Optimization Problems. to appear.

[9] SOKOŁOWSKI, J. Differential stability of solutions to constrained optimization problems. *Appl. Math. Optim.* 13, (1985), 97–115.

[10] SOKOŁOWSKI, J. Sensitivity analysis of control constrained optimal control problems for distributed parameter systems. *SIAM J. Control and Optimization* 25, (1987), 1542-1556.

[11] SOKOŁOWSKI, J. Shape sensitivity analysis of boundary optimal control problems for parabolic systems. *SIAM Journal on Control and Optimization* 26, (1988), 763-787.

[12] SOKOŁOWSKI, J. Stability of solutions to shape optimization problems. *to appear.*

[13] SOKOŁOWSKI, J. AND ZOLESIO, J.P. Shape sensitivity analysis of unilateral problems. *SIAM J. Math. Anal.* 18, (1987), 1416-1437.

[14] SOKOŁOWSKI, J. AND ZOLESIO, J.P. *Introduction to Shape Optimization. Shape sensitivity analysis.* to appear.

[15] ZARANTONELLO, F.H. Projections on convex sets in Hilbert space and spectral theory. *In: Contributions to Nonlinear Functional Analysis, Publ. No.27,Math. Res. Center. Univ.* Wisconsin, Madison, Academic Press, New York , (1971), 237-424.

[16] ZIEMER, P.W. *Weakly Differentiable Functions.* Springer Verlag, New York, 1989.

Murali Rao
Department of Mathematics
University of Florida
201 Walker Hall, Gainesville, FL 32611
USA

Jan Sokołowski
Systems Research Institute
Polish Academy of Sciences
ul. Newelska 6 , 01-447 Warszawa
Poland

J C W ROGERS
# The control of jets at the upstream boundary

Mathematical descriptions of the stability of flows of incompressible fluids in a channel traditionally have been fraught with an element of paradox. On the one hand, there have been studies based on the original ansatz of Rayleigh [1] that perturbations on a basic flow may be represented, in a linear approximation, as a superposition of terms proportional to $e^{i\alpha z}$ where z measures distance along the channel axis and $\alpha$ is real. When the basic flow is steady, one finds that the temporal behavior of the coefficient of $e^{i\alpha z}$ of the form $e^{-i\beta t}$ where $\beta = \beta(\alpha)$ is obtained by solving an eigenvalue problem. In general, $\beta = \beta_1 + i\beta_2$ is not real, and when $\beta_2 > O$, one concludes that the basic flow is not stable, because of the presence of perturbations which grow in time.

On the other hand, real channel flows, even unstable ones, do not behave this way. Instead, perturbations on the flows which are introduced upstream tend to grow in the downstream direction, but remain uniformly bounded in time at upstream locations. An attempt to describe such flows has been based on the following modification of the Rayleigh argument: $\beta$ is real and $\alpha = \alpha_1 + i\alpha_2 = \alpha(\beta)$ may be complex [2,3]. When $\alpha_2 < 0$, the flow is deemed to be unstable.

In any case, it should be possible to describe an evolving flow in terms of an initial- and boundary-value problem. Whether one analyzes the flow in a velocity-pressure or in a vorticity-stream function context, one finds that the instantaneous propagation of signals in the incompressible flow is a consequence of the presence of a Poisson-type equation which holds throughout the flow domain. Typical boundary conditions for such equations entail, in addition to conditions imposed at the sides, the specification of either Dirichlet or Neumann data at the upstream and downstream ends of the flow region. However, the use of Dirichlet or Neumann data at the downstream end carries with it the possibility, when disturbances grow too rapidly downstream, that the flow upstream will be dominated by the details of data imposed in a region where the flow will appear turbulent, and even by the choice of the location of the downstream end.

The way out of this dilemma has been found to involve the imposition of non-standard boundary conditions at the upstream and downstream ends of the region [4]. These boundary conditions lead to well-posed problems, but have the property of propagating and enhancing waves in the downstream direction only. With a little oversimplification, we may describe the essential features of such non-standard boundary conditions for a Poisson-type equation for a function $\psi$ as follows: One of the functions, $\psi$ or $\psi_n$ (the normal derivative of $\psi$), is specified fully at the upstream end, and for the other function a certain number $N$ of "Fourier components" in the "spanwise" direction are prescribed there. To avoid overdetermining $\psi$, the function which, in a standard problem, would be specified at the downstream end, is not prescribed with regard to its corresponding $N$ Fourier components in the spanwise direction.

With these boundary conditions, the problem of the evolution of perturbations on a basic flow may be treated satisfactorily. Nevertheless, some difficulties have remained. The first has concerned the choice of the number $N$. As $N$ is increased, perturbations tend to grow less rapidly, and then to become damped, in time, while their growth in space is enhanced. Hence there should be an optimal, $N = N^*$ such that there is no growth in time, but neither are perturbations forced to be damped excessively in time, at the expense of too rapid a growth downstream. This optimal $N^*$ depends upon the maximum rate of growth of $-\alpha_2(\beta)$ perturbations in the downstream direction and on the geometry of the channel. However, it is unnatural to have to solve a linear boundary-value problem in order to find $N*$, and thus the boundary conditions to be imposed in a proper formulation of the problem.

The second difficulty has been the absence of any indication of a procedure whereby one might achieve solutions of these non-standard boundary-value problems in practice. By way of contrast, the Dirichlet and Neumann boundary conditions can be interpreted as pertaining to systems in contact with reservoirs or to systems which are insulated at their boundaries, respectively. In accordance with these interpretations, solutions of the more usual boundary-value problems satisfy maximum principles with respect to the boundary data. That is not the case with solutions of the non-standard boundary-value problems.

Both of these difficulties have been removed in the case of flows in a channel, and such problems are now being treated numerically [5]. Our purpose here is to show how the same approaches may be applied to the free-boundary analog of channel flow, which is flow of a jet. The stability of a jet was first investigated by Rayleigh [6], and more recently the spatial growth of disturbances has been treated by Keller, Rubinow, and Tu [7].

The unperturbed flow is given by the velocity field $U\mathbf{k}$ $0 \leq r \leq a$, $z \geq 0$ where $r$ is the cylindrical radial coordinate and $z$ is in the axial downstream direction. The undisturbed free surface is at $r = a$. For the perturbed flow, we shall restrict our attention to axially symmetric perturbations and keep only terms which are no more than linear in the disturbance. Let the pressure be $p + \sigma/a$ where $\sigma$ is the coefficient of surface tension, and let the free surface be given by

$$r = a + \xi(z, t) \tag{1}$$

Then at the free surface we have

$$p(a, z, t) = -\frac{\sigma}{a^2}\xi - \sigma\xi_{zz} \tag{2}$$

and for $0 \leq r \leq a$

$$\Delta p = 0 \tag{3}$$

Let the radial component of velocity be $u$ and the axial component be $U + w$. The kinematic free surface condition is

$$\xi_t + U\xi_z = u(a, z, t) \tag{4}$$

In the case of axially symmetric flow, there are three quantities which are candidates for possible prescription at the upstream boundary $z = 0$. They are $u, w$, and $\omega$, the azimuthal component of vorticity. To prescribe all three would lead to a Cauchy problem for the stream function, which would be unstable. Instead, one can obtain well-posed problems by specifying either $w$ and $u$, or $w$ and $\omega$, at $z = 0$. In accordance with the treatment developed for flow in a channel [5], one should prescribe $w$ and $u$ at $z = 0$ for some of the time, and $w$ and $\omega$ at $z = 0$ for the rest of the time. Since the vorticity satisfies

$$\omega_t + U\omega_z = 0 \tag{5}$$

the vorticity is always bounded by its values at $z = 0$ (or $t = 0$). One finds then that, to a good approximation, the flow is irrotational and the perturbation may be derived from a velocity potential $\phi$:

$$u = \phi_r, \quad w = \phi_z \tag{6}$$

In this simplified regime, one can specify either $\phi_r$ or $\phi_z$ at $z = 0$, but not both.

With regard to the matter of controlling the jet through the judicious application of upstream boundary conditions, it would be too much to expect to be able to do this for the whole semi-infinite region $z \geq 0$, especially when there is spatial growth and the flow becomes unbounded as $z \to \infty$ in the linear approximation. Hence we shall consider the control of the flow in a region $0 \leq z \leq L, 0 \leq r \leq a$ where $L$ may be large compared to $a$, but is still finite.

In terms of the velocity potential, the boundary conditions at $r = a$ become, on account of the Bernoulli equation,

$$(\frac{\partial}{\partial t} + U\frac{\partial}{\partial z})^2\phi(a, z, t) = \frac{\sigma}{\rho}(\frac{1}{a^2} + \frac{\partial^2}{\partial z^2})\phi_r(a, z, t) \tag{7}$$

and of course

$$\Delta\phi = 0, \ 0 \leq r < a. \tag{8}$$

Let $\phi(a, z, t)$ be denoted by $f(z, t)$. $\phi_r(a, z, t)$ will depend linearly on $f$ and on the values of $\phi_r$ or $\phi_z$, specified at $z = 0$ and $z = L$, denoted by $\gamma_0(r, t)$ and $\gamma_L(r, t)$, respectively. We denote this dependence symbolically by writing (7) in the form

$$(\frac{\partial}{\partial t} + U\frac{\partial}{\partial z})^2 f = \mathcal{L}f + \mathcal{L}_1\gamma_0 + \mathcal{L}_2\gamma_L \tag{9}$$

According to Rayleigh's theory [6], the operator $\mathcal{L}f$ appearing on the right-hand side of (9) is bounded above by (approximately) $.3432\sigma/(\rho a^3)$ This corresponds to a maximum temporal growth rate $\Omega_0$:

$$\frac{\Omega_0 a}{U} \leq .5858(\frac{\sigma}{\rho a U^2})^{1/2} \tag{10}$$

To treat the stability of the jet as an initial- and boundary value problem, we may choose $N^* \geq$) to satisfy ($\mu_0 = 0$)

$$\mu_{N^*} \leq \frac{\Omega_0}{U} < \mu_{N^*+1} \tag{11}$$

where $J_0(\mu_n a) = 0$, and then prescribe $\phi_z(r, 0, t)$ and the nth Fourier coefficient of $\phi(r, 0, t)$, $int_0^a r J_0(\mu_n r)\phi(r, o, t)dr$ for $1 \leq n \leq N^*$.

Let

$$M = e^{2\mu_{N^*} L} \tag{12}$$

represent a magnification of disturbances in the region $0 \leq z \leq L$. Consider the following solutions of Laplace's equation in the region $0 \leq r < a, 0 < z < L$: Consider the following solutions of Laplace's equation in the region $0 \leq r < a, 0 < z < L$:

$$\Psi^{(i)} = \sum_{k=1}^{i} \Psi^{(k)} \tag{13}$$

$\Psi^{(i)} = 0$ at $r = a$,

$\Psi^{(i)} = 0$ at $z = L$, $i$ odd, $\Psi_z^{(i)} = 0$ at $z = L$, $i$ even,

$\Psi_z^{(i)} = \alpha(r)$ at $z = L$, $i$ odd, $\Psi^{(i)} = \beta(r)$ at $z = L$, $i$ even,

$\Psi^{(2I+1)} = \alpha(r)$ at $z = 0$, and the nth Fourier coefficient of $\Psi^{(2I+1)}(r, 0) - \beta(r)$ is approximately 0 for $n \leq N^*$ and $2I >> M$.

We may use the results of (13) to set up a procedure whereby the evolution of the jet is controlled by the appropriate choice of boundary conditions at the times $m\Delta t$ where $m \geq 0$ is an integer. Denote $\phi(r, z, \Delta t)$ by $\phi^m(r, z)$, and let

$$J\{q\} = \int_0^a r(q(r, 0))^2 \, dr \tag{14}$$

The boundary values $\phi^m(a,z) = f^m(z)$ will be determined from a time-discretization of (9). Given a number $\epsilon > 0$, boundary conditions for $\phi$ at $z = 0$ and $z = L$ will be prescribed as follows:

$$\phi_z^m(r,0) = 0, \quad \phi_r^m(r,L) = 0 \ \ if \ \ J\{\phi_z^{m-1}\} < \epsilon \tag{15a}$$

$$\phi_r^m(r,0) = 0, \quad \phi_z^m(r,L) = \phi_z^{m-1}(r,L) \ \ if \ \ J\{\phi_r^{m-1}\} \geq \epsilon \tag{15b}$$

$$\phi_z^m(r,0) = 0, \quad \phi_r^m(r,L) = \phi_r^{m-1}(r,L) \ \ if \ \ J\{\phi_z^{m-1}\} \geq \epsilon \tag{15c}$$

When $M\Delta t\Omega_0 << 1$, the boundary conditions (15), in tandem with a time-discretized version of (9), will yield the desired control. In the limit as $\Delta t \to 0$, the prescription (15) of boundary conditions will lead to a "bang-bang" sort of control.

References

1 . Lord Rayleigh, Proc. Lond. Math. Soc. **11**, 57 (1880) (Scientific Papers, vol. 1, 474, Dover Publications, 1964)

2. J. Watson, J. Fluid Mech. **14**, 211 (1962)

3. M. Gaster, J. Fluid Mech. **14**, 222 (1962)

4. Joel C. W. Rogers, Quart. Appl. Math. **31**, 199 (1973)

5. Joint work with Robert Handler

6. Lord Rayleigh, Proc. Lond. Math. Soc. 10, 4 (1878) (Scientific Papers, vol. 1, 361, Dover Publications, 1964)

7. Joseph B. Keller, S. I. Rubinow, and Y. 0. Tu, Phys. Fluids **16**, 2052 (1973)

Joel C. W.Rogers
Polytechnic University

# T I SEIDMAN

# Optimal control and well-posedness for a free boundary problem[1]

## 1.  INTRODUCTION

We consider, in radial geometry, a 'crystal grain' of some substance (radius $0 < R = R(t) < L$; concentration normalized to 1) surrounded by a dilute solution with concentration $u = u(t, r)$ in the 'annulus' $R < |x| = r < L$. We consider a fixed time interval $(0, T)$ and set $Q := \{(t, x) : 0 < t < T; R(t) < |x| = r < L\}$. The underlying model involves diffusion[2] in the solution with the concentration satisfying the conservation equation

$$(1.1) \qquad \dot{u} = \Delta u = r^{1-d}(r^{d-1}u_r)_r \quad \text{in } Q$$

with the boundary condition

$$(1.2) \qquad u_r = \dot{R}[1 - u] \quad \text{at the crystal boundary } r = R(t)$$

corresponding to conservation in mass transport 'across' the moving interface. It is convenient to set

$$(1.3) \qquad \dot{R} = h(t) \quad \text{with } R(0) = R_0$$

and this may be viewed as an ordinary differential equation from which to obtain the moving boundary $R(\cdot)$ on $[0, T]$. We also have initial conditions for $u$

$$(1.4) \qquad u(0, r) = u_0(r) \quad \text{for } R_0 < r < L$$

and require, of course, that $0 \le u_0 < 1$ so this represents an admissible concentration. We impose Dirichlet conditions

$$(1.5) \qquad u = \gamma = \gamma(t) \quad \text{at the outer boundary } r = L$$

and will treat the specification of the function $\gamma(\cdot)$ as a *control*; again, we require that $0 \le \gamma \le 1$.

This problem: [the equations (1.1), (1.3) with (1.4), (1.5), and the coupling (1.2)], we will call $(\mathbf{P_0})$. For the analysis, it is convenient to rescale the time-varying $r$-interval $[R, L]$ by setting $y := [L - r]/[L - R]$ with a concommittant alteration of the equation to

$$(1.6) \qquad u_t = \rho^2 u_{yy} - \psi u_y \quad \begin{cases} \rho := \frac{1}{L-R}, \\ \psi := \rho\left[\frac{(d-1)\rho}{\rho-1} + yh\right] \end{cases}$$

[1]This research has been partially supported by the National Science Foundation under the grant ECS-8814788 and by the U.S. Air Force Office of Scientific Research under the grant AFOSR-91-0008.

[2]One could also add a nonlinear reaction term $\varphi(u)$ in the equation. For simplicity we will take $\varphi \equiv 0$, although there are no essential differences to the arguments provided $\varphi$ is differentiable (Lipschitzian) with $\varphi(0) \ge 0 \ge \varphi(1)$.

246

so the problem now 'lives' on the fixed domain $\hat{\mathcal{Q}} := (0,T) \times (0,1)$. We refer to this equivalent rescaled version as $(\hat{\mathbf{P}}_0)$.

Note that, in addition to the initial data $R_0$, $u_0$, the data for $(\mathbf{P}_0)$ or $(\hat{\mathbf{P}}_0)$ consists of the pair of functions $[\gamma, h]$ and we will later, artificially, consider the control problem using this pair. As it stands, $(\mathbf{P}_0)$ is a moving boundary problem but not a *free* boundary problem. We make it a free boundary problem by using the *constitutive relation*

$$(1.7) \qquad\qquad h = H(\kappa, u|_{r=R}) \quad \text{with } \kappa := 1/R$$

to couple the differential equations. We will work with the problem: $[(\mathbf{P}_0) \text{ with } (1.7)]$, which we refer to as $(\mathbf{P}_*)$ with a similar equivalent rescaled form $(\hat{\mathbf{P}}_*)$. Observe that the pair $(1.3)$, $(1.7)$ combine to give

$$(1.8) \qquad\qquad \dot{R} = H(1/R, \omega) \quad \text{with } R(0) = R_0$$

where
$$(1.9) \qquad\qquad \omega := u|_{r=R} \quad (\text{or } \omega := u|_{y=1} \text{ for } (\hat{\mathbf{P}}_*)).$$

and we view $(1.8)$ as an ordinary differential equation for $R$, defining a map

$$(1.10) \qquad\qquad \mathbf{R} : \omega(\cdot) \mapsto R(\cdot) : L^2((0,T) \to [0,1]) \to H^1(0,T)$$

under suitable conditions on $H$, etc. We assume that $H$ is defined[3] and smooth where relevant and that $H(\cdot, 0) \leq 0$. It is not difficult to provide auxiliary structural conditions on the behavior of $H$ for $R \approx 0, L$ which ensure that any solution of the ordinary differential equation $(1.8)$ (for arbitrary $\omega(\cdot)$, taking values in $[0,1]$) remains bounded away from $0, L$ on bounded $t$-intervals. As a simplification, we assume *some* such auxiliary condition and refer to it as $(\mathbf{C})$. This ensures that $(1.8)$ does, indeed, define a map $\mathbf{R}$ as in $(1.10)$ for which $R$ always stays away from $0, L$.

Francis Conrad discussed some aspects of the *steady-state problem* for this at the last of these conferences, at Irsee. Beginning, actually, *during* that conference, Danielle Hilhorst and I then joined him in looking at the time-dependent problem in the setting described above. Let me note here the principal result of that collaboration:

**THEOREM 1:**      *Assume $H$ is as described above, satisfying $(\mathbf{C})$ and with $H(\cdot, 0) \leq 0$. We seek a solution pair $[R, u]$ of*
$(\mathbf{P}_*)$      $[(1.1), (1.3), (1.4), (1.5), (1.2), (1.7)]$
*for $0 < t \leq T$ but shift to the rescaled version $(\hat{\mathbf{P}}_*)$ for convenience. We consider $R(\cdot)$ in, e.g., $H^1(0,T)$ and, consider $u$ in the usual parabolic space*

$$\mathcal{U} := C([0,T] \to L^2(0,1)) \cap L^2((0,T) \to H^1(0,1)).$$

*Let $0 < R_0 < 1$ and let $u_0 \in L^2(R_0, L)$ with $0 \leq u_0 \leq 1$ ae; let $\gamma$ be in $H^1(0,T)$ with $0 \leq \gamma \leq 1$. Then there is a unique solution $[R, u] \in H^1(0,T) \times \mathcal{U}$ and the map:*

---

[3]In real applications the function $H$ typically takes the form: $H(\kappa, \omega) := K[\omega - G(\kappa)]$. We note that the most widely used model is the Ostwald–Freundlich law: $G(\kappa) := C_* e^{\delta \kappa}$, although this cannot be expected to remain valid for very small $R$ ($\kappa \to \infty$). In any case, our results do not depend on any specific form for $H$.

$[R_0, u_0, \gamma] \mapsto [R, u]$ is continuous from $\mathbb{R} \times L^2 \times H^1$ (with the restrictions $0 < R_0 < L$ and $0 \leq u_0, \gamma \leq 1$) to $H^1(0, T) \times \mathcal{U}$ (with the restrictions $0 < R < L$ and $0 \leq u \leq 1$).

PROOF: See [1], [2] but we indicate here the structure of the argument. One begins by considering an artificial problem using truncated functions, modifying $H$, to obtain certain compactness. Later, a Maximum Principle argument shows that the solution obtained is already a solution of the original problem. This compactness then implies existence of a suitable weighted norm on $L^2((0, T) \to [0, 1])$ for which one has strict contractivity of the map $\mathbf{F} = \mathbf{F}_\gamma$ defined by

(1.11) $\qquad \mathbf{F} : \omega \xrightarrow{\quad \mathbf{R} \quad} R \mapsto \dot{R} =: h \xrightarrow{\quad (\hat{\mathbf{P}})_0 \quad} u \xrightarrow{\quad \text{trace} \quad} u|_{y=1}$

where, of course, we are keeping $u_0, \gamma$ fixed in considering $(\hat{\mathbf{P}}_0)$. This gives unique existence and the estimates obtained also give the continuity of the solution map. ∎

## 2.   OPTIMAL CONTROL

A few weeks after the Irsee conference there was a conference on DPS, control of Distributed Parameter Systems, at Santiago de Compostela and I discussed some control-theoretic aspect of the crystal problem, viewing $(\mathbf{P}_*)$ as a *boundary control problem* in two senses: we control to make the free boundary $R(\cdot)$ match some desired evolution and we will be using the boundary data $\gamma$ as the control. Somewhat arbitrarily, we consider a cost functional of the rather standard form

(2.1) $$\mathcal{J} := \mathcal{J}_0(R) + \frac{1}{2} \int_0^T |\gamma - \gamma_*|^2 \, dt$$

where $\gamma_*(\cdot)$ is a convenient 'center' for control and $\mathcal{J}_0$ measures the (cost of) deviation from the desired 'profile'. We will assume only that $\mathcal{J}_0$ is lsc from, e.g., $H^1(0, T)$ and may, for example, take

(2.2) $$\mathcal{J}_0 := \frac{\alpha}{2} \int_0^T |\rho - \rho_*|^2 \, dt$$

with $\rho := 1/(L - R)$ as in (1.6).

The principal difficulty is that $\mathcal{J}$ is not coercive with respect to the $H^1(0, T)$ topology for $\gamma$ for which we have our only existence result, Theorem 1, so we seek to extend that.

THEOREM 2: *Assume the same hypotheses on $H$, $R_0$, $u_0$ as for Theorem 1. Then, for every $\gamma \in \mathcal{G} := L^2((0, T) \to [0, 1])$, there is at least one solution $[R, u]$ of $(\mathbf{P}_*)$ or, equivalently, of $(\hat{\mathbf{P}}_*)$ where $R$ is in $H^1([0, T] \to (0, L))$ as in Theorem 1 and $u$ is in a compact subspace of the Banach space*

$$\mathcal{V}_* := C([0, T] \to L^2(0, 1)) \cap L^2((0, T) \to H^s(\varepsilon, 1)) \qquad (s < 1; \; \varepsilon > 0).$$

*Further, the set of such $[\gamma, R, u]$ is closed in $\mathcal{G} \times H^1(0, T) \times \mathcal{U}_*$, using the weak topology for $\mathcal{G}$.*

PROOF: We cite [4] for details but, as for Theorem 1, indicate the nature of the argument. Here one considers the same map $\mathbf{F}_\gamma$, given as there by (1.11) with a modified $H$. Now one need only argue continuity and compactness of $\mathbf{F}$ to obtain existence

248

through application of the Schauder Theorem since we do not assert uniqueness. For the compactness, the trick is to consider (1.6) and write the solution as $u = v + z$ with

$$z_t = \rho^2 z_{yy}; \qquad z|_{t=0} = u_0, \qquad z|_{y=0} = \gamma, \qquad \rho^2 z_y|_{y=1} = 0.$$

Note that $z$ depends only on $\gamma$ and the further change of time variable to $\tau$ with $d\tau/dt = \rho^2$ enables us easily to verify the regularity of $z$ and show suitable compactness as $\gamma$ ranges over $\mathcal{G}$, taking advantage of 'interior regularity' results. The resulting equation for $v$ now no longer involves $\gamma$ explicitly and $v$ can now be estimated by standard 'energy' methods to obtain applicability of the Aubin Compactness Theorem. ∎

**COROLLARY 2.1:** Let $\mathcal{J}$ be as in (2.1) with $\mathcal{J}_0$ lsc from $H^1(0,T)$ and let $H$ and the data $[R_0, u_0]$ be as in the Theorem. Then $\mathcal{J}$ attains its minimum, i.e., there exists an optimal triple $[\tilde{\gamma}, \tilde{R}, \tilde{u}]$, minimizing $\mathcal{J}$ subject to satisfying $(\mathbf{P}_*)$ in the sense of Theorem 2.

**PROOF:** Consider any minimizing sequence for $\mathcal{J}$: $\{[\gamma_k, R_k, u_k]\}$ satisfying $(\mathbf{P}_*)$. Theorem 2 ensures existence of a convergent subsequence and that the limit $[\tilde{\gamma}, \tilde{R}, \tilde{u}]$ satisfies $(\mathbf{P}_*)$. ∎

## 3. REGULARITY

The results of the last section are somewhat less than satisfactory from the point of view either of partial differential equations or control theory since we do not yet have well-posedness for $\gamma \in \mathcal{G}$. This means that one does not know that $\tilde{\gamma}$ actually *controls*, in the sense of determining the particular optimal solution pair $[\tilde{R}, \tilde{u}]$, and this makes it rather difficult to attempt a characterization of $\tilde{\gamma}$ through (first order) necessary conditions for optimality: we cannot obtain optimality conditions by differentiating $\mathcal{J}$ with respect to its variable if $\mathcal{J}$ is not actually a function of $\gamma$. The trick is to 'decouple' the problem, returning to $(\mathbf{P}_0)$.

**THEOREM 3:** Assume the same hypotheses on $H$, $R_0$, $u_0$ as for Theorems 1,2. Then the map

$$(3.1) \qquad \mathbf{T} : \gamma \times \mathcal{H} \to C([0,T] \to (0,1)) \times \mathcal{G} : [\gamma, h] \xmapsto{\quad (\hat{\mathbf{P}})_0 \quad} [R, \omega]$$

is well-defined and (continuously) Frechet differentiable for a suitable neighborhood $\mathcal{H} \in L^2(0,T)$. The derivative, at each point, is a compact operator.

**PROOF:** This is a fairly straightforward computation, leading to a linear system (which we omit here) coupling a partial differential equation on $\hat{Q}$ with an ordinary differential equation along $y = 1$. See [3]. ∎

The desired well-posedness from $L^2$ is now a simple corollary to this and Theorems 1 and 2.

**COROLLARY 3.1:** The problem $(\mathbf{P}_*)$ is well-posed in the context of Theorem 2.

**PROOF:**    From Theorem 2 we have existence and note a compact set of $h = H(1/R, \omega)$ arising; we take $H$ to be a neighborhood of this set so $\mathbf{R}$ is well-defined. From Theorem 3 we see that the related map $\mathbf{D} : [\gamma, h] \mapsto [h - H(1/R, \omega)]$ is also continuously differentiable and $\mathbf{D}_h$ has the form $\mathbf{I}+[compact]$. One easily verifies from the system giving the derivative that $\mathbf{D}_h$ is injective so, by standard spectral theory, it is boundedly invertible. Application of the Implicit Function Theorem then gives local existence of a $C^1$ map: $\gamma \mapsto h$ for which $\mathbf{D}(\gamma, h) \equiv 0$ — just corresponding to the use of (1.7) as a constraint, converting $(\mathbf{P_0})$ to $(\mathbf{P_*})$. We thus have a *local* well-posedness for $(\mathbf{P_*})$. On the other hand, since $H^1$ is dense in $\mathcal{G}$, we may use Theorem 1 to assert well-posedness globally. ∎

For the control problem we note another corollary.

**COROLLARY 3.2:**    *Assume, for (2.1), that $\gamma_*$ is in $H^1(0, T)$ and that $\mathcal{J}_0$ is 'smooth', say given by (2.2). Then the optimal $\tilde{\gamma}$ of Corollary 2.1 is actually in $H^1(0, T)$.*

**PROOF:**    See [3]. From Theorem 3 and Corollary 3.1 we note that $\mathcal{J}$ is a differentiable function of $\gamma$ and optimality gives $\langle \mathcal{J}'(\tilde{\gamma}), \delta \rangle \leq 0$ for all admissible variations $\delta$, i.e., such that $\tilde{\gamma} + s\delta$ continues to take values in $[0, 1]$. It follows that $\tilde{\gamma}$ is the $[0, 1]$-truncation of $\gamma_* + \Upsilon$ where $\Upsilon$ is obtained through the use of the adjoint of $\mathbf{F}'$, from which we can show that $\Upsilon$ is in $H^1$. ∎

# References

[1] F. Conrad, D. Hilhorst, and T. Seidman, *On a reaction-diffusion equation with a moving boundary* in *Recent Advances in Nonlinear Elliptic and Parabolic Problems*, (P. Benilan, M. Chipot, L.C. Evans, M. Pierre, eds.), Pitman, to appear.

[2] F. Conrad, D. Hilhorst, and T. Seidman *Well-posedness of a moving boundary problem arising in a dissolution–growth process*, Nonlinear Anal-TMA, to appear.

[3] P. Neittaanmäki and T. Seidman, *Optimal solutions for a free boundary problem for crystal growth*, in *Control of Distributed Parameter Systems*, (F. Kappel, K. Kunisch, W. Schappacher, eds.) Springer Lect. Notes Control, Inf. Sci., to appear.

[4] T. Seidman, *Some control-theoretic questions for a free boundary problem* in *Control of Partial Differential Equations*, (LNCIS #114, A. Bermúdez, ed.), Springer-Verlag, New York, pp. 265–276, (1989).

**Thomas I. Seidman**
Department of Mathematics and Statistics
University of Maryland Baltimore County, Baltimore, MD 21228, USA

# Electromagnetism and electronics

A BOSSAVIT
# A free boundary problem in magnetostatics

**Abstract**. Magnetostatics consists in finding vector fields $b$ and $h$ such that $\operatorname{div} b = 0$, $\operatorname{rot} h = j$ ($j$ is given), and $b = \Gamma(h)$, where $\Gamma$ is the subgradient of a convex functional $U$ (the magnetic coenergy). When $U$ is the sum of a quadratic functional and of a support functional, a 2-phase Stefan problem results, a "vector" Stefan problem so to speak, because the unknown is a vector-field, not a function, like for instance the temperature.

## Introduction

What follows is a short presentation of the problem, with the limited purpose of explaining how a free boundary appears in the context of some questions of magnetostatics, and of emphasizing the "vectorial" character of this Stefan-like problem. For a study of the underlying mathematical structure and a numerical approach with mixed elements, please refer to [3].

Consider the following situation (Fig. 1). A current coil $C$ drives a magnetic flux through a torus $M$, made of magnetizable material (mild iron). What is the relation between this flux and the total current, or *characteristic* of the circuit (to be defined with more precision below), and how is the magnetic field distributed?

The relevant equations are:

$$(1) \qquad \operatorname{rot} h = j, \qquad\qquad (2) \quad b = \Gamma(h), \qquad\qquad (3) \quad \operatorname{div} b = 0,$$

where $b$ and $h$ (vector fields, unknown) are the induction field and the magnetic field, and $j$ (a vector field, given) is the current density in $C$. The crucial feature is the form of the non-linear behavior law $b = \Gamma(h)$: at point $x$, the value of the field $b$ is

$$(4) \qquad b(x) = \gamma_x(|h(x)|)\ h(x)/|h(x)|,$$

with $\gamma_x$ a maximal monotone graph as in Fig. 2. Due to the nature of this non-linearity, there are two regions in $M$, separated by a free boundary: the saturated region, where $|b(x)| > b_0(x)$ (cf. Fig. 2) and the non-saturated one, where $h(x) = 0$.

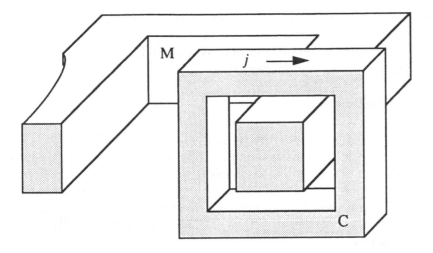

**Figure 1**. The problem. Only one half of the system is shown. The other half is symmetrical with respect to the vertical plane. Currents in the coil C drive a magnetic flux in (the whole space but mainly in) the magnetic circuit M.

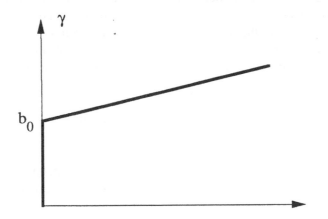

**Figure 2**. Graph of $\gamma$, a maximal monotone graph of $IR^+ \times IR^+$ (not $IR \times IR$). The slope being $\mu_0$ ($4\pi\ 10^{-7}$ in MKSA units), $\gamma$ is the subdifferential of the function $f: \eta \rightarrow b_0\ \eta + \mu_0\ \eta^2/2$. The "saturation threshold" $b_0$ is 0 in the air and may depend on position in the steel region (it depends on several factors, especially the temperature).

The $\gamma$ of Fig. 2 is an idealization of the real b—h curve, which has a steep section near the origin, not a true discontinuity. The point of this "simplification" (which in fact rather complicates matters from a strictly numerical point of view) is to generate a Stefan-like two-phase problem where the unknown field is *vectorial*, not scalar.

Only the mathematical modelling will be done here. The model thus obtained (eqs.(6) to (10) below) has a strong mathematical structure, including some "complementarity" properties (in the sense familiar to students of convex analysis; cf. e.g. [2, 6]). For this, and for its finite-elements solution, we refer to [3].

## Mathematical modelling

Starting from $\gamma$, let us define $f : IR \rightarrow IR$ by $f(\eta) = \int_0^\eta \gamma(s)\,ds$ and let $g(\beta) = \sup\{\eta \in IR : \beta\eta - f(\eta)\}$ be its Fenchel conjugate. The inequality $f(\eta) + g(\beta) \geq \eta\beta$ holds, with equality iff the pair $\{\eta, \beta\}$ lies in the graph of $\gamma$. Given a vector field $h$ [resp. $b$] over $IR^3$, let us set $U(h) = \int_{IR^3} f(|h(x)|)\,dx$ and $V(b) = \int_{IR^3} g(|b(x)|)\,dx$. Then $U$ and $V$ are convex, lower semi-continuous (l.s.c.) functionals, in duality. In the case under consideration, $U(h) = \int_{IR^3} (b_0(x)\,|h(x)| + \mu_0\,|h(x)|^2/2)\,dx$, and $V(b) = \mu_0^{-1}\int_{IR^3} ((|b(x)| - b_0)^+)^2/2\,dx$. The equality

$$(5) \qquad U(h) + V(b) = \int_{IR^3} h(x) \cdot b(x)\,dx$$

means that relation (4) holds a.e. in $x$. (Note that if $h$ and $b$ are unrelated, the *inequality* $\geq$ holds in (5).) We may (according to an old idea, [5]) take advantage of this to consider other behavior laws than (4): give a pair $U$ and $V$ of convex l.s.c. functions in duality on $IL^2(IR^3)$, then instead of (1)(2)(3) set the more general problem of finding $h$ and $b$ such that (1), (3) and (5) hold. What will be said below applies to this general situation. Note that anisotropy and non-homogeneity of the magnetic material may easily be taken into account this way, but not hysteresis.

As a function of $h$ [resp. $b$], $U$ [resp. $V$] is called the "magnetic coenergy" [resp. "magnetic energy"].

So the problem is: given $j$ (with $\text{div } j = 0$), *find* a pair $\{h, b\}$ of vector fields over $IR^3$ *such that* (1), (3) and (5) *hold*. But leaving it in such form would be ignoring an essential aspect of the situation, which is physically intuitive, and could be established via an asymptotic analysis [4]: the field concentrates in $M$, and the air around contributes only a very small part of its total energy. So a computation which aims at finding the distribution of the field in $M$ may ignore what happens in the air, and thus may deal with (1), (3) (5) in $M$ only. By using symmetry, one then faces a boundary-value problem in the domain $D$ of Fig. 3 (half of $M$). Now what about boundary conditions on the surface $S$ of $D$? On the part $S_h$ which belongs to the symmetry plane, $n \times h = 0$ by an easy symmetry argument. On the remaining part $S_b$, $n \cdot b = 0$, because the flux-lines of field $b$ stay inside $M$. The initial problem has become: *find* $\{h, b\}$ *in* $IL^2_{\text{rot}}(D) \times IL^2_{\text{div}}(D)$ *such that*

$$(6) \qquad \text{rot } h = 0 \text{ in } D, \quad n \times h = 0 \text{ on } S_h,$$

$$(7) \qquad \text{div } b = 0 \text{ in } D, \quad n \cdot b = 0 \text{ on } S_b,$$

(8)    $U(h) + V(b) = \int_D h \cdot b,$

where U and V are as above, but defined as integrals over D, not over the whole space.

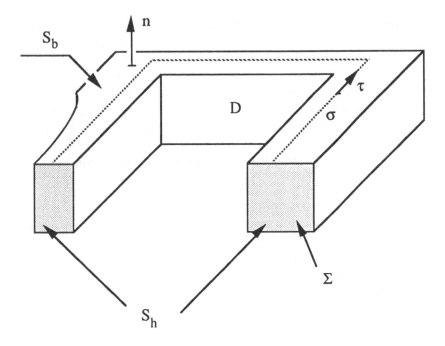

**Figure 3.** The problem-domain, with its boundary S in two parts ($S_h$ and $S_b$). The normal n, of length 1, is outward directed at all regular points of S, and $\tau$ is the unit tangent vector along path $\sigma$. The "cutting surface" $\Sigma$ (which here is one of the connected components of $S_h$, but could be placed differently, provided it "cuts" the magnetic circuit) is in a kind of dual topological relationship with $\sigma$ (cf. [1], p. 53).

What about the source of the field (which previously was in the right-hand side of eq. (1))? It now appears as a linear constraint on the circulation of field h along any path linking the two connected components of $S_h$, like $\sigma$ on Fig. 3:

(9)    $\int_\sigma \tau \cdot h = J,$

where J is half the total intensity (intensity in the thread multiplied by half the number of turns) in the coil of Fig. 1. Note that (9) is a consequence of Stokes' theorem (known in electromagnetism as "Ampère's theorem") when applied to (1), and is valid for all paths similar to $\sigma$. It can be shown that problem (6)(7)(8)(9) is well-posed (at least as far as h is concerned; b can obviously be non-unique in the case of Fig. 2).

Condition (9) privileges h, because J is the source of the field in the present situation, but mathematically speaking a similar condition on b, that is

(10) $\qquad \int_\Sigma n \cdot b = F,$

could be substituted for it (F is the flux of induction sustained by the magnetic circuit). This would make sense if the aim of the modelling was to answer the question: "which J should be applied in order to get the flux F ?". One would then solve (6)(7)(8)(10), and obtain J by evaluating the line integral (9). The symmetry between these two formulations suggests a more general approach, which answers an enlarged family of questions: find the *pairs* {J, F} of values which are compatible with (6, 7, 8). Taken all together, they form a graph in the J—F plane (whose right part is similar in shape to the graph of Fig. 2), which is called the *characteristic curve* of the circuit. In [B], we show how to obtain some information about the location of this graph in the J—F plane: *it lies inside a set which can be explicitly constructed* (computed).

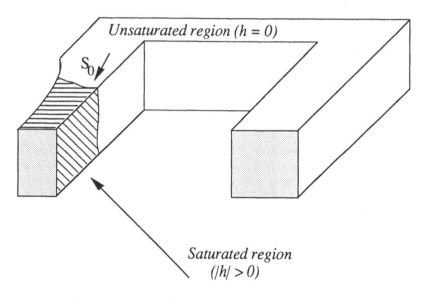

**Figure 4.** Conjecture about the free boundary: it cuts the circuit, and intersects its surface at right angles.

## The free boundary

About the question of the free boundary, I believe (but could not prove) that for positive, not too large values of J, the situation is as in Fig. 4. The difficulty lies in establishing the *smoothness* of the free boundary $S_0$. For under this assumption, n x h = 0 on $S_0$, and thus b (parallel to h in the saturated region) is normal to $S_0$. Therefore, the flux of b through $S_0$

is equal to its area times $b_0$, and $S_0$ cuts the surface $S_b$ of the circuit at right angles (since n . b = 0 on $S_b$). So for a certain range of values of F (those such that $F/b_0$ is intermediate between the smaller and the larger cross-section area of the circuit), we can predict the situation of Fig. 4. Note that $S_0$ must cut (disconnect) the domain because of Ampère's theorem. (In the language of algebraic topology, $S_0$ is in the same homology class (relative to $S_b$) as the $\Sigma$ of Fig. 3.)

One should not forget, in discussing this point, that the mathematical model has been obtained by going to the limit in two different ways: first by selecting the shape of $\gamma$, then by assuming that no flux goes through $S_b$. In real magnetic circuits, flux lines tend to avoid saturated regions and so may take a shortcut through the air, in contradiction with the previous assumption. The present model is valid if the dimensionless ratio, or *Stefan number* of the problem, $Lb_0/\mu_0 J$ (where L is the length of the path $\sigma$ of Fig. 3), is large enough, but not *too* large: large enough, because when it tends to infinity, the limit boundary condition is indeed n . b = 0 on $S_b$, but not too large, because the "squaring off" of the b—h curve we did in Fig. 2 could then no longer be justified.

# References

[1]  P. Alexandrov: **Elementary concepts of topology**, Dover (New York), 1961. (First published in 1932, Springer.)

[2]  G. Allen: "Variational Inequalities, Complementary Problems and Duality Theory", **JMAA**, 58 (1979), pp. 1-10.

[3]  A. Bossavit: "Mixed elements and a two-phase Stefan problem in magnetostatics", in **Proc. Conference on Numerical Methods for Free Boundary Problems** (P. Neitaanmäki, ed.), July 23-27, 1990 in Jyväskylä, Finland, **Int. series of Numerical Mathematics, Vol. 99,** Birjhaüser Verlag (Basel), 1991, pp. 93-102.

[4]  A. Bossavit: "On the Condition 'h Normal to the Wall' in Magnetic Field Problems", **Int. J. Numer. Meth. Engng., 24** (1987), pp. 1541-50.

[5]  J.J. Moreau: "Applications of convex analysis to the treatment of elastoplastic systems", in **Applications of methods of Functional Analysis to problems in Mechanics,** Symp. IUTAM-IMU, Lecture Notes in Math. 503 (P. Germain, B. Nayroles, eds.), Springer (Berlin), 1976.

[6]  J.T. Oden, J.N. Reddy: "On Dual Complementary Variational Principles in Mathematical Physics", **Int. J. Engng. Sci.,** 12 (1974), pp. 1-29.

A. Bossavit
Électricité de France, 1 Av. du Gal de Gaulle,
92141 Clamart

W MERZ
# Oxidation of Silicon

## 1. Introduction

We introduce a one dimensional moving boundary value problem describing the oxidation of silicon. Two free boundaries are involved, the interface $Si$–$SiO_2$ and the outer surface $SiO_2$–$O_2$. The growth of silicon dioxide is a thermal process which takes place at high temperatures. The model presented in this paper assumes a temperature of $1000^0$ C and the $SiO_2$ is regarded as a compressible, viscous fluid.

Several simultaneous phenomena occur: The oxidizing species diffuse throughout the oxide and on the $Si$–$SiO_2$ iterface, the oxygen reacts with the silicon forming silicon dioxide. The chemical reaction causes a movement of this interface into the silicon. The difference in the densities of $Si$ and $SiO_2$ cause an overall expansion of the oxide as well as the motion of the outer surface into the oxygen region. (See Fig. 1)

The resulting fluid dynamical model includes a set of partial differential equations, the Diffusion–, the Navier–Stokes– and the Continuity Equations. Appropriate boundary conditions are deduced, taking into account "sharp" interfaces.

We describe the numerical simulation of the oxidation process and present numerical results.

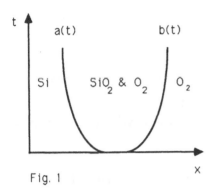

Fig. 1

## 2. Assumptions and Hypotheses

Description of the physical quantities used in the model:

$\tilde{\varrho}$ : density of silicon.

$C^*$ : equilibrium concentration of oxygen within the dioxide.

$p^*$ : air pressure.

$g$ : gravitational constant.

For $SiO_2$:

$\varrho$ : density of $SiO_2$.

$v$ : velocity of the fluid, i. e. $SiO_2$.

$p$ : pressure.

$\sigma$ : stress in the dioxide.

For $O_2$:

$C$ : concentration of the oxygen diffusing throughout the oxide.

$D$ : coefficient of diffusion.

$h$ : transition coefficient of $O_2$ into the $SiO_2$.

$k$ : transaction coefficient between $O_2$ and $Si$.

259

Additional quantities, assumed to be valid in the $SiO_2$ region are:

$\lambda$ and $\mu$ are viscosities with $3\lambda + 2\mu = 0$, thus

$$\sigma = -p + \lambda\, v_x + \mu\, v_x = -p + \frac{1}{3}\,\mu\, v_x$$

Throughout this paper we denote the partial derivatives of a function by subscripts $x$ and $t$, representing the space– and the time derivative, respectively. A dot over a time dependent function means its time derivative.

We assume the following properties:

1. $Si$ is a rigid body and its density $\tilde{\varrho}$ is constant.
2. $SiO_2$ is a compressible, viscous fluid and $\varrho = \alpha p + \beta$, where $\alpha \geq 0$, $\beta > 0$ are constants.
3. The temperature is constant ($1000^0$ C).
4. The viscosity depends on the temperature only and is therefore a constant.
5. $O_2$ doesn't influence the mechanical properties of $SiO_2$.
6. The flow of the oxidizing species is given by $q = -DC_x + vC$.
7. $Si + O_2 = SiO_2$ describes an irreversible chemical reaction of first order, where no heat is released.

## 2. The Mathematical Model

In order to describe the complete oxidation process, we have to consider the diffusion of the oxidizing $O_2$ molecules, the motion of the fluid and the conservation of mass. Moreover, we have to describe the chemical reaction at the interface $a$ and the motion of the outer interface $b$.

On $\Omega(t) = \{(x,t)|a(t) < x < b(t), 0 < t < T_f\}$, $T_f > 0$ we get the following partial differential equations:

$$C_t(x,t) = DC_{xx}(x,t) - \big(v(x,t)\, C(x,t)\big)_x, \tag{1}$$

$$\varrho(x,t)\,\big[v_t(x,t) + v(x,t)\, v_x(x,t)\big] = -p_x(x,t) + \frac{\mu}{3}\, v_{xx}(x,t) - g\,\varrho(x,t), \tag{2}$$

$$\varrho_t(x,t) + \big(\varrho(x,t)\, v(x,t)\big)_x = 0, \tag{3}$$

$$\varrho(x,t) = \alpha\, p(x,t) + \beta, \tag{4}$$

with appropriate initial conditions $C(x,0) = C_0(x)$, $v(x,0) = v_0(x)$ and $\varrho(x,0) = \varrho_0(x)$.

The boundary conditions for equation (1) are of the following form:

$$DC_x(a(t),t) - v(a(t),t)\,C(a(t),t) = k\,C(a(t),t) \quad \text{on } a(t)\,, \tag{5}$$

$$-DC_x(b(t),t) + v(b(t),t)\,C(b(t),t) = h(C(b(t),t) - C^*) \quad \text{on } b(t)\,. \tag{6}$$

Condition (5) describes the chemical reaction.

For equation (2) we use

$$\sigma = -p^*$$

as a boundary condition on $b$, i. e.

$$-p(b(t),t) + \frac{\mu}{3}\,v_x(b(t),t) = -p^* \quad \text{on } b(t)\,. \tag{7}$$

In order to derive the remaining boundary conditions for equation (2) and (3) as well as equations describing the movement of the free boundaries, we consider some rates of change:

$$V_1 = \varrho(b(t),t)\,\dot b(t) - \varrho(a(t),t)\,\dot a(t) - \varrho(b(t),t)\,v(b(t),t) + \varrho(a(t),t)\,v(a(t),t)\,,$$
$$V_2 = -\tilde\varrho\,\dot a(t)\,,$$
$$V_3 = k\,C(a(t),t)\,,$$
$$V_4 = \varrho(a(t),t)\,v(a(t),t)\,,$$

where

  $V_1$ denotes the rate at which $SiO_2$ is produced,
  $V_2$ is the rate at which $Si$ is consumed,
  $V_3$ represents the rate at which $O_2$ is consumed and
  $V_4$ denotes the rate at which molecules flow into the $SiO_2$ at $a$.

We observe that $V_1 = V_2 = V_3 = V_4$. Completing the set of boundary conditions, we use for equation (2)

$$v(a(t),t) = -\frac{\tilde\varrho}{\varrho(a(t),t)}\,\dot a(t) \quad \text{on } a(t)\,. \tag{8}$$

Since the velocity $v$ is positive, the Continuity equation requires one condition on the left hand–side boundary,

$$\varrho(a(t),t) = \frac{\dot{b}(t) - v(b(t),t)}{\dot{a}(t) - v(a(t),t)}\, \varrho(b(t),t) + \frac{\tilde{\varrho}}{\dot{a}(t) - v(a(t),t)}\, \dot{a}(t) \quad \text{on } a(t). \quad (9)$$

Finally, the displacement of the free interfaces $a$ and $b$ is given by two ordinary differential equations of the form

$$\dot{a}(t) = -\frac{k}{\tilde{\varrho}}\, C(a(t),t), \quad (10)$$

$$\dot{b}(t) = \frac{\varrho(a(t),t)}{\varrho(b(t),t)}\, \dot{a}(t) + v(b(t),t), \quad (11)$$

with appropriate initial conditions $a(0) = a_0$ and $b(0) = b_0$, $b_0 - a_0 > 0$.

Together with the initial conditions, equations (1)–(11) describe the oxidation problem completely.

3. The Numerical Simulation

In order to solve the oxidation problem numerically, we discretize the $SiO_2$ region as shown in Fig. 2. The space mesh is variable at each time step, such that the free boundaries $a$ and $b$ always pass through a grid point.

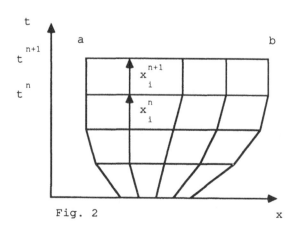

Fig. 2

Let $\Delta t$ be the time step and let $m > 0$ be an integer. We consider the mesh points $(x_i^n, t^n)$, where $t^n = n\Delta t$, $n > 0$ and $x_i^n = a(t^n) + i\Delta x^n$, $\Delta x^n = \frac{b(t^n) - a(t^n)}{m}$, $0 \leq i \leq m$.

We determine the approximate solution of the oxidation problem at successive time levels by using *fully implicit* finite difference methods. The approximated values of $C(x_i^n, t^n)$, $v(x_i^n, t^n)$ and $\varrho(x_i^n, t^n)$ are denoted by $C_i^n$, $v_i^n$ and $\varrho_i^n$, respectively, where $0 \leq i \leq m$ and $n > 0$. Applying equations (10) and (11), we can predict the locations of the free interfaces $a(t)$ and $b(t)$ at the next time level. If we track the boundaries implicitly, we have to iterate at each time step. Let $a^n$ and $b^n$ be the approximations of $a(t^n)$ and $b(t^n)$, respectively. Suppose the computations at the time level $t^n$ have been completed and we want to compute the values at time $t^{n+1}$. Using implicit schemes for the approximation of the ordinary differential equations (10) and (11), we get:

$$a^{n+1} = -\Delta t\left(\frac{k}{\tilde{\varrho}} C_0^{n+1}\right) + a^n,$$

$$b^{n+1} = \Delta t\left(\frac{\varrho_0^{n+1}}{\varrho_m^{n+1}} \dot{a}^{n+1} + v_m^{n+1}\right) + b^n,$$

where the subscripts $0$ and $m$ denote the values of our functions on the boundaries $a$ and $b$, respectively.

At this point, an iteration method must be used. Let $k > 1$ be an integer and let $a^{n+1,k}$, $b^{n+1,k}$, $\{C_i^{n+1,k}\}_{0 \leq i \leq m}$, $\{v_i^{n+1,k}\}_{0 \leq i \leq m}$ and $\{\varrho_i^{n+1,k}\}_{0 \leq i \leq m}$ denote the approximations obtained after the kth iteration, then we compute:

$$a^{n+1,k+1} = -\Delta t\left(\frac{k}{\tilde{\varrho}} C_0^{n+1,k}\right) + a^n,$$

$$b^{n+1,k+1} = \Delta t\left(\frac{\varrho_0^{n+1,k}}{\varrho_m^{n+1,k}} \dot{a}^{n+1,k+1} + v_m^{n+1,k}\right) + b^n.$$

$C_i^{n+1,k}$, $v_i^{n+1,k}$, $\varrho_i^{n+1,k}$ are known values and $C_i^{n+1,0} = C_i^n$, $v_i^{n+1,0} = v_i^n$, $\varrho_i^{n+1,0} = \varrho_i^n$ for $0 \leq i \leq m$ and $n > 0$.

Using the values $a^{n+1}$ and $b^{n+1}$, we can compute the new step size $\Delta x^{n+1}$ of the space grid and we are now able to solve the oxidation problem at the time level $t^{n+1}$.

Therefore, let's define the quantity

$$\dot{s}_i^{n+1} = \frac{x_i^{n+1} - x_i^n}{\Delta t}$$

for $0 \leq i \leq m$ and $n > 0$.

We approximate the time derivatives of equation (1) and (2) at the grid point $(x_i^{n+1}, t^{n+1})$ by the following implicit scheme:

$$\frac{C_i^{n+1} - C_i^n}{\Delta t} \simeq C_t(x_i^{n+1}, t^{n+1}) + \dot{s}_i^{n+1} C_x(x_i^{n+1}, t^{n+1}),$$

$$\frac{v_i^{n+1} - v_i^n}{\Delta t} \simeq v_t(x_i^{n+1}, t^{n+1}) + \dot{s}_i^{n+1} v_x(x_i^{n+1}, t^{n+1}),$$

where the "numerical convection terms" $\dot{s}_i^{n+1} C_x$ and $\dot{s}_i^{n+1} v_x$ take into account the slopes of the grid with respect to time.

The derivatives $C_x$, $v_x$ and $\varrho_x$ at the point $(x_i^{n+1}, t^{n+1})$ are approximated by *backward differences*, i. e.

$$C_x \simeq \frac{C_i^{n+1} - C_{i-1}^{n+1}}{\Delta x^{n+1}},$$

$$v_x \simeq \frac{v_i^{n+1} - v_{i-1}^{n+1}}{\Delta x^{n+1}},$$

$$\varrho_x \simeq \frac{\varrho_i^{n+1} - \varrho_{i-1}^{n+1}}{\Delta x^{n+1}}.$$

We use the typical space centered scheme for the second spatial derivative of the functions in equation (1) and (2).

Finally, we have to consider the non–linear term $vv_x$, contained in the Navier–Stokes equation. We approximate it by an iterative scheme of the following form:

$$v(x_i^{n+1}, t^{n+1}) v_x(x_i^{n+1}, t^{n+1}) \simeq v_i^{n+1,k} \frac{v_i^{n+1,k+1} - v_{i-1}^{n+1,k+1}}{\Delta x^{n+1,k+1}}$$

$$+ v_i^{n+1,k+1} \frac{v_i^{n+1,k} - v_{i-1}^{n+1,k}}{\Delta x^{n+1,k}} - v_i^{n+1,k} \frac{v_i^{n+1,k} - v_{i-1}^{n+1,k}}{\Delta x^{n+1,k}},$$

for $k = 0, 1, 2, \cdots$. Notice that $v_i^{n+1,0} = v_i^n$ and $\Delta x^{n+1,0} = \Delta x^n$.

264

The discretization of the initial conditions and the boundary conditions (8) and (9) is obvious. We approximate the boundary conditions (5), (6) and (7) by the second order scheme:

$$C_x(a(t^{n+1}), t^{n+1}) \simeq \frac{C_{-1}^{n+1} - C_1^{n+1}}{2\Delta x^{n+1}} \ ,$$

$$C_x(b(t^{n+1}), t^{n+1}) \simeq \frac{C_{m-1}^{n+1} - C_{m+1}^{n+1}}{2\Delta x^{n+1}} \ ,$$

$$v_x(b(t^{n+1}), t^{n+1}) \simeq \frac{v_{m-1}^{n+1} - v_{m+1}^{n+1}}{2\Delta x^{n+1}} \ .$$

This completes the numerical considerations of the oxidation problem and the discrete equations can be derived easily.

## 4. Numerical Results

Most of the physical data used for the numerical evaluation were obtained from [1] and [2]. The initial value of the density of the $SiO_2$ was set to be 2.224 $gr/cm^3$. The portion of the density between $SiO_2$ and $Si$ was assumed to be 0.44, see [2]. The viscosity $\mu = 1.02 \times 10^{13}$ poise. We compute the oxidation process at an air pressure of 760 Torr. In this work all quantities are expressed in *number of molecules*. The initial value of the density for instance, in terms molecules becomes $\varrho_0 = 22.2918$ *molecules/nm$^3$*.

Fig. 3 shows the typical profile of the free boundaries $a$ and $b$. We used the time step $\Delta t = 0.05$ throughout the computation, 8–15 grid points on the space net, depending on the growth of the oxide thickness and 6 iteration steps at each time level.

265

Fig. 4 shows the oxygen distribution within the $SiO_2$ at the last time level $t = 4$ min. At the boundary $a$ it is almost zero, at the boundary $b$ we get $C(b(t), t) = C^*$ , i. e. $0 \leq C \leq C^*$ during the whole oxidation process and the concentration is slightly decreasing in time on the remaining $SiO_2$ region. The concentration $C^* = 5.2 \times 10^{-5}$. The values on the vertical axis are multiplied by $10^6$.

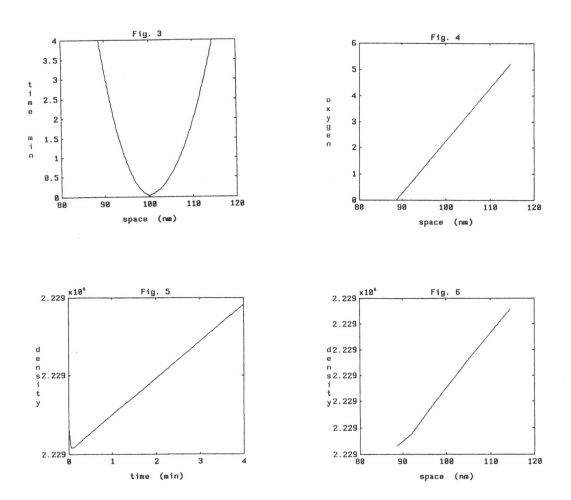

In Fig. 5 we see the time dependence of the density of silicon dioxide at the boundary $a$ and Fig. 6 presents the space dependence of $SiO_2$ at the last time level. It increases slightly in space as well as in time.

Fig. 7 represents the oxide thickness at an oxidation time of half an hour.

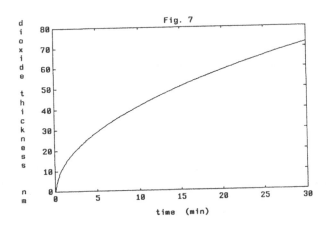

5. References

[1] B. E. Deal and A. S. Grove, General Relationship for the Thermal Oxidation of Silicon, *J. Appl. Phys.*, Vol. 36, pp. 3770–3778, 1965.

[2] M. J. Needs, V. Jovic, J. B. Waddell, C. Taylor, K. Board and M. J. Cooke, Numerical Modelling of Local Oxidation of Silicon using the Finite Element Method, Dept. of Civil Engineering, University College, Singleton Park, Swansea, U. K., Dept. of Electrical Eng., University College, Singleton Park, Swansea.

[3] H. Matsumoto and M. Fukuma, Numerical Modelling of Non–Uniform Si Thermal Oxidation, *IEEE Transactions on Electron Devices*, Vol. Ed–32, No. 2, Feb. 1985.

[4] C. Cuvelier, A. Segal and A. A. van Steenhoven, Finite Element Methods and Navier–Stokes Equations, D. Reidel Publishing Company, Dordrecht/Boston/Lancester/Tokyo 1985.

[5] R. B. Guenther and J. W. Lee, Partial Differential Equations of Mathematical Physics and Integral Equations, Prentice–Hall Englewood Cliffs, New Jersey 1988.

W MERZ Oregon State University Department of Mathematics
Kidder Hall 368 Corvallis, Oregon 97331–4605 U.S.A.

C SCHMEISER

# Free boundaries in semiconductor devices*

**Abstract**—The first analyses of semiconductor devices in the 1940s and 50s already considered a splitting of devices into regions of zero space charge and depletion regions. Here it is shown that these simplified models appear as singular limits of the drift-diffusion model for charge transport in semiconductors. The limiting problem for the potential is similar to the so called space charge problem. For $pn$-diodes under certain biasing conditions, an existence result is a by-product of the present analysis.

## 1. Introduction

A nondimensionalized and simplified version of the classical drift-diffusion model for a steady flow of electrons and holes in a semiconductor is given by the elliptic system

$$\lambda^2 \Delta V = n - p - C,$$
$$\text{div}(\gamma \nabla n - n \nabla V) = 0, \tag{1}$$
$$\text{div}(\gamma \nabla p + p \nabla V) = 0$$

for the unknowns $V, n$ and $p$ denoting the electrostatic potential, the electron density and the hole density, respectively. The equations hold in a domain $\Omega \subset R^3$ representing a $pn$-diode. A $pn$-diode is characterized by the fact that the (given) doping profile $C(\mathbf{x})$, $\mathbf{x} \in \Omega$, is positive in a subdomain $\Omega_n$ of $\Omega$ and negative in another subdomain $\Omega_p$ with $\overline{\Omega}_n \cup \overline{\Omega}_p = \overline{\Omega}$. These subdomains are called $n$-region and $p$-region, respectively. The dimensionless parameters

$$\lambda = \frac{1}{L}\sqrt{\frac{\varepsilon U_{bi}}{qC_{max}}}, \quad \gamma = \left(\log \frac{C_{max}}{n_i}\right)^{-1}$$

result from the scaling where the maximal doping concentration $C_{max}$, the built-in voltage $U_{bi} = U_T/\gamma$ and the device diameter $L$ have been used as reference quantities for densities, the potential and length, respectively. The constants $\varepsilon, q, n_i$ and $U_T$ are the permittivity, the elementary charge, the intrinsic density and the thermal voltage, respectively. The scaling is standard (see [5]) with the exception that, to the author's knowledge, the built-in voltage has not been used for the scaling of the potential before.

The boundary of the device consists of two Ohmic contacts, $C_n \subset \partial\Omega \cap \partial\Omega_n$ and $C_p \subset \partial\Omega \cap \partial\Omega_p$, and an insulating part $\partial\Omega_N = \partial\Omega \setminus (C_n \cup C_p)$. At the contacts, the carrier densities satisfy the conditions of zero space charge and thermal equilibrium:

$$n - p - C = 0, \quad np = e^{-2/\gamma}, \quad \text{on } C_n \cup C_p \tag{2a}$$

---

* This work has been supported by the National Science Foundation under Grant No. DMS–890813 and by the Österreichischer Fonds zur Förderung der wissenschaftlichen Forschung under Grant No. J0397–PHY.

which can be rewritten as Dirichlet conditions for $n$ and $p$. The contact potential is given by

$$V = U + 1 + \gamma \log \frac{C + \sqrt{C^2 + 4e^{-2/\gamma}}}{2}, \quad \text{on } C_n,$$

$$V = -1 - \gamma \log \frac{-C + \sqrt{C^2 + 4e^{-2/\gamma}}}{2}, \quad \text{on } C_p$$

(2b)

where $U$ is the applied voltage. The densities as well as the potential satisfy homogeneous Neumann conditions along the insulating boundary parts. See [5] for more details on the model.

In many practical situations the parameters $\lambda$ and $\gamma$ are small compared to one. This motivates considering the limits of (1), (2) as these parameters tend to zero. The limit $\lambda \to 0$ has received considerable attention (see [4] and [5] and the references therein). On the other hand, for voltages of the order of magnitude of $\lambda^{-2}$, a limiting problem similar to that for $\gamma \to 0$ has also been considered ([1], [3], [6], [7], [8]). In this work, $\lambda$ is kept fixed, and the limit $\gamma \to 0$ is carried out under the assumption

$$0 \le U < 2$$

on the contact voltage. An analogous analysis can be carried out for $-2 < U < 0$ but is omitted here for lack of space.

In the following it will be convenient to use the so called quasi Fermi potentials $\phi_n, \phi_p$ which are related to the potential and the densities by

$$n = \exp\left(\frac{V - \phi_n - 1}{\gamma}\right), \quad p = \exp\left(\frac{\phi_p - V - 1}{\gamma}\right).$$

These new variables satisfy the formally self-adjoint differential equations

$$\text{div}(n\nabla\phi_n) = 0, \quad \text{div}(p\nabla\phi_p) = 0$$

and the boundary conditions

$$\phi_n = \phi_p = U, \quad \text{on } C_n, \qquad \phi_n = \phi_p = 0, \quad \text{on } C_p.$$

A proof of the following existence result can be found in [4].

**Theorem 1.** *The problem (1), (2) has a solution which satisfies*

$$0 \le \phi_n, \phi_p \le U,$$

$$-1 - c\gamma \le V \le U + 1 + c\gamma$$

*where here and in the following $c$ denotes positive constants independent of $\gamma$.*

For our purposes the value of the estimates in the theorem is limited because they do not imply uniform estimates for the charge carrier densities. Section 2 contains a collection of stronger estimates which are then used in Section 3 for the convergence analysis.

## 2. A Priori Estimates

**Lemma 1.** *The derivatives of the quasi Fermi potentials satisfy*

$$\|\nabla \phi_n\|_{L^2(\{V > \phi_n + U - 1 + 2\varepsilon\})} \leq c \exp\left(-\frac{\varepsilon}{2\gamma}\right),$$

$$\|\nabla \phi_p\|_{L^2(\{V < \phi_p + 1 - U - 2\varepsilon\})} \leq c \exp\left(-\frac{\varepsilon}{2\gamma}\right)$$

*for every positive $\varepsilon \leq 2 - U$.*

*Outline of a proof.* In the nonempty set $\Omega_0 = \{\phi_p - 1 \leq V \leq \phi_n + U - 1 + \varepsilon\}$, the densities $n$ and $p$ are uniformly bounded. Therefore also $\Delta V$ is bounded by the Poisson equation. This can be used to show that $\Omega_0$ is big enough such that there exists a function $\phi_D$ satisfying the Dirichlet boundary conditions for $\phi_n$, such that $\nabla \phi_D$ vanishes in $\{V > \phi_n + U - 1 + \varepsilon\}$ and $\phi_D$ is uniformly bounded in $H^1(\Omega)$. The estimate for $\phi_n$ of the lemma then follows from the energy estimate

$$\| \exp\left(\frac{V - \phi_n - U + 1 - 2\varepsilon}{2\gamma}\right) \nabla \phi_n\|_{L^2(\Omega)} \leq \| \exp\left(\frac{V - \phi_n - U + 1 - 2\varepsilon}{2\gamma}\right) \nabla \phi_D\|_{L^2(\Omega)}.$$

The second inequality is shown similarly.#

**Lemma 2.** *The densities $n$ and $p$ are uniformly bounded in $L^\infty(\Omega)$.*

*Outline of a proof.* Let $S$ denote a component of $\{V > \phi_n + 1\}$. Then the potential satisfies either $V = \phi_n + 1$ or a homogeneous Neumann condition along $\partial S$. Also,

$$\phi_n = c(\gamma) + O\left(\exp\left(\frac{U - 2}{4\gamma}\right)\right), \quad \text{in } S$$

holds with some $\gamma$-dependent constant $c$ by the previous lemma (with $\varepsilon = \frac{2-U}{2}$). The maximum principle gives

$$V \leq c(\gamma) + 1 + c_1 \gamma \leq \phi_n + 1 + c_2 \gamma$$

implying the boundedness of $n$ in $S$ and therefore in $\Omega$. The boundedness of $p$ is shown analogously.#

A consequence of lemma 2 is the uniform boundedness of $\Delta V$. For the following result we need regularity properties of the boundary. We assume that the boundedness of the densities implies the boundedness of the potential in $H^2(\Omega)$, that extensions of the Dirichlet data as $H^2(\Omega)$-functions exist and that the boundary is a union of segments where the usual trace theorems can be applied.

**Lemma 3.** *Under the above assumptions we have*

$$\gamma\|\nabla n\|_{L^2(\Omega)} = \|n(\nabla V - \nabla\phi_n)\|_{L^2(\Omega)} \le c\sqrt{\gamma},$$
$$\gamma\|\nabla p\|_{L^2(\Omega)} = \|p(\nabla\phi_p - \nabla V)\|_{L^2(\Omega)} \le c\sqrt{\gamma}.$$

*Proof.* Let an $H^2(\Omega)$-extension of the Dirichlet data for $n$ be denoted by $n_D$. Multiplication of the electron continuity equation by $n - n_D$ and integration by parts gives

$$\gamma\int_\Omega |\nabla n|^2 dx = \gamma\int_\Omega \nabla n.\nabla n_D \, dx + \int_\Omega n\nabla V.(\nabla n - \nabla n_D)dx$$

$$= -\gamma\int_\Omega n\Delta n_D \, dx + \gamma\int_{\partial\Omega} n\frac{\partial n_D}{\partial\nu}ds - \int_\Omega \frac{n^2}{2}\Delta V \, dx$$

$$+ \int_{\partial\Omega} \frac{n^2}{2}\frac{\partial V}{\partial\nu}ds - \int_\Omega n\nabla V.\nabla n_D \, dx \le c$$

where $\nu$ denotes the outward unit normal along $\partial\Omega$. The first estimate in the lemma follows immediately, and the second one can be shown similarly.#

## 3. The limiting problem

The results of the previous section imply that for $\gamma \to 0$ every sequence of solutions has a subsequence such that $V$ converges to $V_0$ in $C^1(\overline{\Omega'})$ with $\Omega' \subset\subset \Omega$, and $n, p, \phi_n$ and $\phi_p$ converge to $n_0, p_0, \phi_{n0}$ and $\phi_{p0}$, respectively, in $L^\infty(\Omega)$-weak *. Our estimates permit passing to the limit in the Poisson equation and in the boundary conditions for $V$. Therefore,

$$\lambda^2\Delta V_0 = n_0 - p_0 - C,$$

$$V_0 = U + 1, \quad \text{on } C_n, \qquad V_0 = -1, \quad \text{on } C_p, \qquad \frac{\partial V_0}{\partial\nu} = 0, \quad \text{on } \partial\Omega_N$$

holds. Now we multiply the electron continuity equation by a test function $\varphi \in H_0^1(\Omega \cup \partial\Omega_N)$ and integrate by parts:

$$\int_\Omega \nabla\varphi.(\gamma\nabla n - n\nabla V)dx = 0$$

Lemma 3 implies that the diffusion term tends to zero as $\gamma \to 0$. From an analogous argument for the hole continuity equation we obtain

$$\text{div}(n_0\nabla V_0) = 0, \quad \text{div}(p_0\nabla V_0) = 0.$$

By eliminating the charge carrier densities we arrive at the third order differential equation

$$\text{div}((\lambda^2\Delta V_0 + C)\nabla V_0) = 0$$

for the potential. This is a generalization of the equation with $C = 0$ appearing in the so called space charge problem [2]. The difference is the presence of a fixed background

271

charge density. So far we have proven the solvability of this equation subject to the limiting boundary conditions for the potential. However, it is not clear under what circumstances the zero space charge boundary condition

$$\Delta V = 0, \quad \text{on } C_n \cup C_p$$

holds in the limit. The situation is similar to that for the space charge problem where some results are known for special cases [2].

Consider a $\gamma$-independent open set $S$ in $\Omega$ where the electron density $n$ is uniformly bounded away from zero. Lemma 1 implies that $\nabla \phi_n$ tends to zero in $S$. On the other hand, the difference between $\nabla V$ and $\nabla \phi_n$ vanishes in the limit by lemma 3. Therefore, $\nabla V_0 = 0$ holds in $S$. It is easy to see that $p_0 = 0$ holds in $S$. From the limiting Poisson equation we, thus, obtain $n_0 = C$ in $S$. Since, obviously, $n_0$ is nonnegative such a set $S$ has to be a subset of the $n$-region $\Omega_n$. In the same way it can be shown that sets where the limiting hole density $p_0$ is nonnegative lie in the $p$-region, that $\nabla V_0$ and $n_0$ vanish there and that $p_0 = -C$ holds. On the other hand, if $n_0 = p_0 = 0$ holds in an open set, then the limiting potential is a solution of $\lambda^2 \Delta V_0 + C = 0$. This suggests that the potential actually solves the equation

$$(\lambda^2 \Delta V_0 + C) \nabla V_0 = 0$$

where in the regions with $\nabla V_0 = 0$ we have zero space charge ($n_0 - p_0 - C = 0$), and the regions with $\lambda^2 \Delta V_0 + C = 0$ are depletion regions ($n_0 = p_0 = 0$). Recalling that we have $V_0 \in C^1(\Omega)$ and $-1 \leq V_0 \leq U + 1$ it is obvious that a solution of the limiting problem can be obtained by solving a standard obstacle problem. (For a discussion of its asymptotic behaviour as $\lambda \to 0$ see [8].) However, in general this is not the only solution. Finding additional information which completes a well posed problem for the limiting potential remains the subject of further study.

## References

[1] F. Brezzi, A. Capelo, and L. Gastaldi, A singular perturbation analysis of reverse biased semiconductor diodes, *SIAM J. Math. Anal.* **20** (1989), pp. 372–387.

[2] C.J. Budd, A. Friedman, B. McLeod, and A.A. Wheeler, The space charge problem, IMA preprint No. 454, Minneapolis, Minnesota, 1988.

[3] L. Caffarelli and A. Friedman, A singular perturbation problem in semiconductors, *Bolletino U.M.I.* **1–B 7** (1987), pp. 409–421.

[4] P.A. Markowich, *The Stationary Semiconductor Device Equations*, Springer-Verlag, Wien—New York, 1986.

[5] P.A. Markowich, C. Ringhofer, and C. Schmeiser, *Semiconductor Equations*, Springer-Verlag, Wien—New York, 1990.

[6] R.E. O'Malley and C. Schmeiser, The asymptotic solution of a semiconductor device problem involving reverse bias, to appear in *SIAM J. Appl. Math.* **50** (1990).

[7] C. Schmeiser, On strongly reverse biased semiconductor diodes, *SIAM J. Appl. Math.* **49** (1989), pp. 1734–1748.

[8] C. Schmeiser, A singular perturbation analysis of reverse biased pn-junctions, to appear in *SIAM J. Math. Anal.* **21** (1990).

**Christian Schmeiser**  Department of Mathematical Sciences
Rensselaer Polytechnic Institute Troy, NY 12180–3590, USA

Printed and bound by CPI Group (UK) Ltd, Croydon, CR0 4YY

21/10/2024

01777093-0012